The Electric Vehicle
and the Burden of History

THE ELECTRIC VEHICLE
AND THE BURDEN
OF HISTORY

DAVID A. KIRSCH

RUTGERS UNIVERSITY PRESS
New Brunswick, New Jersey, and London

Library of Congress Cataloging-in-Publication Data

Kirsch, David, 1964–
    The electric vehicle and the burden of history / David A. Kirsch.
      p. cm.
    Includes bibliographical references and index.
    ISBN 0–8135–2808–9 (cloth : alk. paper) — ISBN 0–8135–2809–7 (paper : alk paper)
    1. Automobiles, Electdric—History. 2. Automobiles—Design and construction—
History. 3. Automobile industry and trade—United States—History. I. Title.
TL220.K57 2000
629.22'93'09—dc21

                                                            99–056542

British Cataloging-in-Publication data for this book is available from the British Library

Excerpt from "Burnt Norton" in FOUR QUARTETS by T. S. Eliot, copyright 1936 by Harcourt, Inc., and renewed 1964 by T. S. Eliot, reprinted by permission of the publisher.

Manufactured in the United States of America

To Andrea

Contents

Figures and Tables

TABLES

Acknowledgments

If the electric vehicle has struggled through the twentieth century under cumulative disadvantage, this book—and its author—have benefited from extraordinary cumulative advantages, and it is my distinct pleasure to acknowledge a small subset of those who have helped me bring this book to completion.

I would like to thank William Perry and Nathan Rosenberg for bringing me to Stanford, where the research for the doctoral dissertation on which this book is based began, and for introducing me to my principal adviser, Timothy Lenoir. Over the years, Tim never wavered in his support of my work. It was also a privilege to work with other distinguished scholars such as Paul David and Donald Kennedy. My work has profited greatly from close readings by people who have thought deeply about issues of science, technology, and government. Gabrielle Hecht was a constructive critic and helped me navigate the shoals of dissertation writing. Other members of the Stanford community also offered valuable guidance and input: Joe Corn, Paul Edwards, Michael Hannan, Henry Lowood, and Ed Steinmuller either reviewed drafts or helped me clarify my thinking about various aspects of the larger project.

Since moving to the Anderson School at UCLA in 1996, I have been welcomed and supported by many new colleagues. For their broad interest in integrating business history and the history of technology with the study and teaching of management, I would especially like to thank Steven Lippman, Richard Rumelt, and Chairman and later Interim Dean John Mamer. At various points, Sushil Bickchandani, Lyle Brenner, Charles Corbett, Chris Erickson, Sandy Jacoby, David Lewin, and Ben Zuckerman also offered much appreciated advice and encouragement.

Among my fellow historians, Clay McShane and Michael Schiffer have been unfailingly generous with their time and energy. Beyond their personal interest in me and my work, their prior research has materially improved this work in ways too many to count. I have also benefited from the personal and intellectual generosity of Gijs Mom, whose parallel work on the history of the electric vehicle should be seen as an essential complement to the perspectives that follow.

Many others have helped guide my research. Among them, Jesse Ausubel, director of the Program for the Human Environment at Rockefeller University, has been a constant source of inspiration. Arnulf Grübler of the International Institute for Applied Systems Analysis first encouraged me to investigate the problem of technological choice in the history of the automobile; he and Nebojša Nakičenovič offered valuable reactions to early drafts of this work. At the Maastricht Economic Research Institute on Innovation and Technology (MERIT), Christopher Freeman challenged me to think about connecting work in the economics of technological change to my interest in environmental policy, and René Kemp helped me think through many of the issues discussed in chapter 8.

On the practical side, I would like to thank Julie Popkin of Popkin Literary Agency for helping me turn the dissertation into a book and for introducing me to Helen Hsu at Rutgers University Press and to Monica Faulkner in Los Angeles. During my final months of work, Monica helped me clarify the structure and logic of the argument of the book and even found time to laugh at one or two of my jokes. I have also benefited from the dedicated and capable research assistance of Shauna Mulvihill and Nancy Hsieh, who helped gather and organize many of the materials that follow. In addition, I received generous support from the Alfred P. Sloan Foundation, the AT&T Foundation, and the Andrew W. Mellon Foundation.

On a more personal note, Andrea Nagin Kirsch has stood by my side for most of the past fifteen years and helped me see this project through from start to finish. An acknowledgment is truly inadequate to the task. As I entered the writing home stretch, my gestating family was thrust into the hands of Dr. Khalil Tabsh and the nursing staff of the Santa Monica Hospital. I do not know how to express my gratitude to them for delivering me two beautiful babies and returning Andrea to me whole, intact, indeed perfect. In the spirit of this study, Isabel's and Jacob's future choices are real thanks to their dedicated and skillful care.

Finally, thanks are due to my mother and stepfather, Marguerite and Josh, my first and finest teachers.

All mistakes are mine.

Santa Monica, August 1999

PART ONE
CONTEXT

What might have been and what has been
Point to one end, which is always present.
Footfalls echo in the memory
Down the passage which we did not take
Towards the door we never opened
Into the rose-garden. My words echo
Thus, in your mind.
                T. S. Eliot, "Burnt Norton," *Four Quartets*

David A. Kirsch

The Electric Vehicle and the Burden of History

Errata

The second sentence on page 3 contains an error. The sentence should read: "From actor-activist Ed Begley, Jr., to *Baywatch* actress Alexandra Paul, the first leaseholders of General Motors' EV1 paid a stiff price to lead the vanguard of automotive technology." *Tonight Show* host Jay Leno owns a 1909 Baker electric vehicle, but does not lease an EV1.

INTRODUCTION

On an overcast December morning in Los Angeles in 1996, a handful of celebrities took delivery of the first mass-produced electric cars made and sold by a major American auto maker since before World War I. From *Tonight Show* host Jay Leno to *Baywatch* actress Alexandra Paul, the first leaseholders of General Motors' EV1 paid a stiff price to lead the vanguard of automotive technology. After the journalists, photographers, and gawkers got back into their gasoline-powered cars and headed home, each of the lucky few electric vehicle drivers was left with a $35,000 two-seater with limited range and a checkered past. To the untrained eye, these high-priced roadsters and their glamorous occupants hardly suggest that an automotive revolution is at hand; a second, third, or fourth car for Hollywood bigwigs seems unlikely to wean Americans of their penchant for petroleum. How can this fledgling market ever hope to transform the way in which Americans move from place to place? Will electric cars be affordable and fun enough to coax Americans out of their internal combustion–powered cars? Will electric cars eventually play an important part in the national system of motor transport? Will electric vehicles help liberate the American economy from excess dependence on petroleum? Given the pervasive presence of the automobile, answers to these questions are crucial to understanding the future course of American society.

This book sets out to answer these questions by looking at the history of the electric vehicle in the United States. In the late 1890s, at the dawn of the automobile era, steam, gasoline, and electric cars all competed to become the dominant automotive technology. By the early 1900s the battle was over, and internal combustion was poised to become the prime mover of the twentieth century. For many people the idea that a number of the nation's first cars were powered by electricity is surprising. Others familiar with bits and pieces of American automotive lore see electric cars as bygone relics of technological failure. Like the fabled Stanley Steamers, electric cars are viewed as technological dead ends, historical footnotes to the storied rise of Henry Ford and the rest of the American automobile industry. Although most automotive historians have argued that internal combustion was destined to become the prevailing standard by virtue of its inherent technological superiority, these accounts put the cart before the horse. Rather than accepting the technological superiority of internal combustion as a given, I contend that such superiority cannot be determined without social context—in other words, technological superiority was ultimately located in the hearts and minds of engineers, consumers, and drivers, not programmed inexorably into the chemical bonds of refined petroleum. From this perspective, the choice of internal combustion takes on added dimensions, reflecting social and cultural relations in late nineteenth-century American society; and this period of technological uncertainty emerges as a crucial turning point in the history of American transportation technology.

Understanding this process of technological choice is also important because the initial competition to power the automobile—like that between VHS and Betamax, eight-track tapes and cassettes, IBM personal computers and Apple Macintoshes—established a durable standard, in this instance one that has lasted nearly a century. What characteristics of late nineteenth-century American society led to the choice of internal combustion over its steam and electric competitors? Might other factors, under slightly differing initial conditions, have led to the adoption of another motive power as the technological standard for the American automobile?

In this book I make three essential claims about the relationship among the gasoline engine, the automobile, and twentieth-century American society. First, the builders of the American automobile industry did not and could not have anticipated that their decisions would definitively shape the evolution of the American transportation system in the twentieth century. The earliest manufacturers and consumers could not have envisioned the fantastic success of the self-powered motor vehicle; the pervasive spread of the automobile surprised even its most enthusiastic supporters. Many contemporaries expected that a range of technological variants would survive the process of technological stabilization, resulting in a permanently shared market among some combination of steam, gasoline, and electric vehicles. Given that this early period was a time of great technological variation, there was no prima facie reason to expect that variability to disappear. In other words, internal combustion emerged as the dominant technology only after the fact, as a result of the path dependent evolution of the expanding automobile market; yet that dominance continues to shape transportation choices and policies a full century later.

Second, other alternative developmental pathways were available. The automobile was much more than a simple artifact; it organized an entire system of transportation technologies. The four-wheeled, transmission-linked, privately owned and operated internal combustion–powered vehicle was not the only possible technology. Because the heart of the stand-alone, motorized road transport system was being negotiated at this time, other systemic variants were also available. Functionally equivalent transportation systems might have been organized around third-party service providers—for example, building on traditional institutions such as the livery stable and the hansom cab. Although most people do not think of it as such, the standard internal combustion vehicle that we have inherited is a specific type of technological hybrid that incorporates important attributes of both electric and internal combustion designs. This hybrid character usually goes unnoticed. It is quite possible, however, that technological hybridization might have resulted in different patterns of functional specialization: electric cars might have continued to provide urban transport service (perhaps as "city cars"), with gasoline and steam cars used for touring and long-distance travel. In short, there were many possible answers to the question "what is an automobile?" and the "electrified gas car" that we have come to know was only one such answer.

Third, the history of the American automobile is one instance of the recurrent dynamic interaction between technological choice and long-run environmental sustainability. Instead of accepting current claims of an impending disaster if we continue to rely upon internal combustion, I argue that there is no such thing as an environmentally friendly automotive technology. The first automobiles, whether gasoline-, steam-, or electric-powered, represented a dramatic environmental improvement over the horse-drawn technology that they replaced. The social, financial, and environmental threats we now face as a result of our reliance on refined petroleum are not the fault of internal combustion technology per se but of the massive expansion of the automobile transport system. In this sense, the scale of the automotive technology system is its most salient and potentially hazardous characteristic. Had steam or electricity succeeded in establishing an alternate technological standard and had such a system been in place today, we would still face serious (although different) social and environmental costs. The dynamic relationship between technological and environmental change will continue into the future. The high-roller, electric vehicle purchasers of the late 1990s (should they prove to be the vanguard of a new century of automotive technology) may think they are solving one set of environmental problems, but they are contributing to another of yet unknowable dimensions. Ultimately, technological and economic adaptation to social and environmental constraints will begin again, and another episode of revolutionary technological change will ensue.

The literature on the history of the automobile is almost as vast and sprawling as the built environment to which the private passenger car has given rise. In this book I focus on episodes within the larger history, presenting a series of commentaries on the history of the modern motor car. Each chapter describes a different component of the automotive system—itself a complex of cultural values, infrastructure networks, historic patterns of circulation and exchange, and technological artifacts. The vignettes offer windows into the larger historical narrative and point to major issues in the broader phenomena of motorization and technological competition. Several little-known historical episodes are covered in considerable detail, while other well-documented events are noted only as they affected the process of technological selection.

Chapter 1 establishes the intellectual and methodological context for reexamining the history of the electric car. Compressing the historiography of the automobile into a few short pages, the chapter places the central organizing concept of path dependent technological choice within the larger setting of the evolving transportation system. Traditional writing about the history of the automobile has focused too slavishly upon the artifact—the black box of the car it-self. Viewed in this light, the electric vehicle inevitably suffers. Using the disparate literatures on automotive, electrical, and urban trans-port history, chapter 1 attempts to draw a more representative pic-ture of the electric vehicle as part of the larger transportation system.

Part II, including chapters 2 through 5, presents the historical narratives at the heart of this study. Chapter 2 covers the history of the Electric Vehicle Company (EVC), providing a starting point for a vision of an alternative automotive transport system in which gaso-line and electric vehicles would have been used to supply different kinds of transport service. For fifteen years, from its founding in 1897 as the Electric Carriage and Wagon Company until 1912 when its descendant, the New York Transportation Company, ceased electric operations, the company operated electric vehicles for hire in and around New York City. Prior accounts of this episode have focused either on battery limitations, financial chicanery by the company's founders and owners, or the link between the EVC and the Selden patent. Yet the company was a landmark venture in the history of American transportation. The moxie of its founders, the scope of their vision—to own and operate fleets of electric taxicabs across the United States and throughout the world—begs for further inquiry. The com-plex tale of the company's rise and fall captures the spirit of the era. For a brief period, the electric vehicle was the dominant technology, supported by electrical manufacturers, streetcar operators, and other vested interests. Significantly, however, the EVC embodied a differ-ent view of the role of the horseless carriage in which passengers paid for mobility and service rather than buying, maintaining, and oper-ating their own vehicles. By 1902 the failure of the company's na-tional ambitions, amid swirling accusations of mismanagement and fraud, prefigured the collapse of the service-based view of the future of the automobile. Thereafter, the company's activities were limited to New York City, where as many as six hundred of its electric cabs continued in service until 1912. In short, chapter 2 tells the story not just of the EVC but of the alternative, urban, service-based transpor-tation system that might have been.

Chapter 3 takes up the issue of organizational support for elec-tric vehicles. Although the electric utility industry acted too late to influence the standardization of the American automobile, several progressive light and power companies recognized the potential value of the vehicle charging market. Leading electric companies (in Boston,

Hartford, New York, and Philadelphia) used electric vehicles to re-place horse-drawn vehicles for lamp lighting, delivery, and other station-related duties. Not until 1909, however, did electric vehicle supporters create the Electric Vehicle and Central Station Association, an organization expressly dedicated to encouraging electric power and light companies to use electric vehicles for their own transportation needs. The following year, Thomas Edison himself gave his blessing to the Electric Vehicle Association of America (EVAA), a full-fledged trade organization representing electric vehicle manufacturers, bat-tery makers, and electric companies. During its six-year existence as an independent entity, the EVAA helped underwrite a modest resur-gence of interest in the electric vehicle, especially for commercial de-livery and haulage. Had the industry been able to organize itself in support of electric vehicles by 1900 instead of 1910, the urban ser-vice market might have become self-sustaining regardless of the ulti-mate fate of the private electric passenger vehicle.

Chapter 4 focuses on the history of the electric truck and explores the demise of the notion of separate spheres—the idea that different forms of transportation technology might coexist, each in its own distinct sphere of service. In stark contrast to the passenger vehicle market, demand for commercial vehicles evolved more slowly. Elec-tric trucks maintained a significant share of the commercial market well into the second decade of the century. Early adopters of all types of motor vehicles suffered from problems with reliability and unpre-dictable operating costs. Merchants were wary of unknown technol-ogy, and change was incremental. Owners of horse fleets gradually experimented with one or more electric- or gasoline-powered trucks, often for marketing and advertising. The new vehicles shared quar-ters and duties with the remaining horses, leading to the emergence of distinct spheres of service that persisted for years. Only as com-mercial motor vehicles spread to more and smaller businesses did the logic of separate spheres begin to unravel, a process that was acceler-ated by World War I and its impact on the American truck industry. Small-scale operators could not afford the multiple vehicles needed to take advantage of functional specialization. Seeking a single, uni-versal truck, local merchants had little alternative but to choose in-ternal combustion. This chapter also describes the battery exchange system implemented at Hartford between 1911 and 1924 whereby truck owners could exchange batteries at will to extend service range and reduce uncertainties associated with battery maintenance. The end of separate spheres ultimately signaled the end of the electric truck.

Chapter 5 steps back to examine the larger context within which steam, internal combustion, and electricity were competing. Turn-of-the-century American infrastructure shaped the choices of turn-of-the-century American motorists. Roads and especially access to fuel and charging facilities set the boundaries within which technological choices had to be made. As in any social process, however, human agents adapted the prevailing infrastructure to suit individual purposes. Touring, the preferred activity of early motorists, was difficult in an electric vehicle; yet many drivers accepted the challenge and met with varying degrees of success. The issue of electric touring exposed a fundamental fissure within the community of electric vehicle supporters. Manufacturers, battery makers, and light and power companies could not agree about how to best position the electric vehicle to enable the industry to survive and grow. Ultimately, the electric vehicle infrastructure could not be all things to all people. Lacking clear direction, and despite the impressive efforts of the electric touring enthusiasts, the brief electric renaissance of the 1910s could not be sustained.

Following a brief excursus on the role of counter-factual logic and the "accidents" of history, Part III presents three different perspectives connecting the historical episodes of Part II with more recent automotive history. The burden of history—expectations and the accumulation of institutional and cultural inertia over the course of automotive history—is the subject of chapter 6. Early electric vehicle enthusiasts had many reasons to hope for a revolutionary breakthrough in energy storage technology; their generation had lived through the last quarter of the nineteenth century, an age of technological miracles. Initially, faith in the imminent solution to the battery problem ran high. Over time, however, hope gave way to a mixture of steadfast optimism and wistful resignation. Expectations were never fulfilled, even as incremental technological changes dramatically improved the capabilities of the typical electric vehicle. All the while, internal combustion was consolidating its hold on the automobile market, further raising the bar for a successful electric passenger vehicle. Gradually, the electric car came to occupy a unique position; its prospects always seemed bright, even though memories were full of its history of unmet expectations.

Shifting the focus from technological competition to the

interaction of electricity and internal combustion, Chapter 7 looks briefly at the history of hybrid vehicles—those combining one or more different drive technologies—and extrapolates from the hybrid vehicle to explore the legacy of hybridization for the history of the automobile transportation system. In this view, the electrification of the automobile emerges as a significant long-run trend despite the evident failure of the stand-alone electric vehicle. Because the chapter considers the electrified gas car (the particular hybrid configuration we know today as the modern automobile) as merely one among several possible endpoints to the early evolution of the automobile industry, it also connects past experiences to present and future prospects for alternative-fuel vehicles. If the hybrid system we depend on today was only one among many possible alternatives, then perhaps there is some basis to hope that a different hybrid configuration will become the standard of the future.

Chapter 8 returns to the issue of technology and environment and the problem of scale. Drawing on recent work in the field of industrial ecology, the final pages offer a framework for understanding why unwelcome social and environmental aspects of large-scale systems of technology are so difficult to anticipate. In the realm of policy, several techniques have been proposed to help identify and mitigate the unintended consequences of technological change. Unfortunately, the legacy of the electric vehicle offers only scant evidence for expecting that we will make better choices in the future.

Can the electric vehicle ever hope to escape the burden of history? Will any combination of technology, marketing, and regulation allow the electric car to catch up to internal combustion? In a word, no. History suggests that current attempts to reintroduce electric vehicles will not succeed. Perhaps the commercial fuel cell vehicle is closer to reality today than all the other promised commercial electric vehicles of the past were in their day. Nothing would please me more than to be proved wrong on this score. My findings, however, suggest that the future of the electric car will not be found in better batteries but in better systems of delivering mobility to future end users. Only if service providers are willing to challenge their—and our—historically and culturally bounded notions of what constitutes an automobile will we ever escape the harsh logic of internal combustion.

RESCUING THE ALTERNATIVES TO INTERNAL COMBUSTION

1

Large Technical Systems, Technological Competition, and the Burden of History

> At that busy corner, Grand Street and the Bowery, there may be seen cars propelled by five different methods of propulsion—by steam, by cable, by underground trolley, by storage battery, and by horses.
>
> *New York Sun*, 1898

Today, we may complain about the complexities of airline pricing policies, long-distance calling plans, computer software, and other late twentieth-century technological accouterments. But the changes confronting late nineteenth-century urbanites must have seemed even more overwhelming. Standing in a quiet doorway observing that "busy corner" in the immigrant center of the nation's fast-changing urban landscape, a New Yorker saw a world palpably in flux and wildly unpredictable. Overhead, the steam-powered locomotives of the New York Elevated belched smoke and ash into the air. Beneath them, the cable cars of the Third Avenue Railroad lurched up and down the Bowery, sharing the intersection with the recently electrified cars of the Metropolitan Street Railway's Madison Avenue line. Interspersed among the tried and true horse cars trudging slowly along the Grand Street crosstown line were a few horseless streetcars powered by electricity drawn from massive lead-acid batteries.

But out on the streets and off the rails were horse carts of every shape and size. From light one-horse traps to fancy carriages to heavy teams hauling thousands of pounds of freight, the streets of New York were awash with horses, urine, manure, and flies. An extremely lucky or patient observer might have noticed a rare horseless carriage. These, too, came in assorted shapes, sizes, and modes of propulsion; but the most common was the electric cab introduced in 1897 by two Philadelphia entrepreneurs in partnership with the Electric Storage Battery

Company. Passing quietly by, riding on their oversized inflatable tires, the horseless carriages looked strangely awkward. Although they bore a strong resemblance to the familiar horse-drawn hansom cabs, which had a uniformed man perched high atop each vehicle controlling its operation with mechanical levers rather than reins, they lacked the visual counterbalance of the horse leading the way and seemed somehow incomplete. Beneath the driver, in the "boot," rested nearly the same lead-acid batteries that powered the crosstown streetcars.

What would the future hold? How would early twentieth-century New Yorkers—and residents in other rapidly expanding cities—get themselves from place to place? The revolutionary claims of their promoters notwithstanding, none of these transportation systems seemed ready to displace the others. The overhead trolley, the one solution that had been spreading throughout American cities for a decade, was not allowed on the crowded streets of Manhattan.

Today, standing on the corner of Grand Street and the Bowery, you will see exactly one prime mover: internal combustion. In New York you might also hear the rumble of the subway, but in most American cities and towns the subway is not an option. Above ground, there remains little evidence of the battles waged a century ago. In New York City and across the United States, a sea of motorcycles, cars, cabs, sport utility vehicles, vans, and trucks testify to the dominance of internal combustion. What happened? How and why did internal combustion, an obscure and all but invisible technology one hundred years ago, become the standard? And what does the fate of the vanquished alternatives tell us about ourselves?

Like many large-scale technological systems, motorized road transport was shaped by complex interactions among existing social and physical structures, actors and agents, and local circumstances. No single person, law, or geological feature was responsible for how the automobile industry developed. The presence of these multiple and overlapping factors meant that the automobile system was subject to path dependence: future choices were shaped by those who chose first.[1] Because the mature, culturally embedded automobile system of today still bears the marks of these earlier episodes, it is reasonable to ask whether an alternative to internal combustion might have succeeded and, if so, what that system might have looked like. Although the steam car briefly dominated the fledgling American motor vehicle industry, the battery electric vehicle emerged as the only viable alternative and then only for urban transport and other niche markets. This alternative hybrid system would have operated very differently from the internal combustion transportation system. In retrospect, given the multivalent role that the automobile has played in shaping American society in the twentieth century, an alternative to internal combustion might have had equally far-reaching implications.

To share this particular historical adventure, one must be willing to make several small but crucial leaps of faith—to look beyond nearly every vehicle one sees today to a time before the automobile and accept the following claims. The automobile was, and is, only one part of the transportation system. The laws of physics and the geography of American cities did not require the automobile to become what it is today. We late twentieth-century residents of developed countries did not simply inherit the automobile. Rather, it was chosen for us approximately one hundred years ago by a select group of system builders and drivers. The transportation system that we inherited, which is based on privately owned and operated internal combustion–powered vehicles, was only one possible solution to the challenge of motorizing transit over public roads. Other alternatives, based upon an expanded and integrated role for semi-private transport, may seem unthinkable today. But to the typical urban American at the turn of the century, our motorized world would appear equally remote and unbelievable. The prospect of a privately owned, mass-produced motor vehicle for every American adult man and woman would have seemed nothing less than lunacy.

No one can know exactly what the alternative to internal combustion would have looked like because, of course, it never came into being. We have only its shadows to guide us—a few days of electric cab operating records here, a lost opportunity there. We can see its traces in the boastful predictions of electrical engineers drunk with the self-confidence of being present at the dawn of the electrical age. Yet these are small pieces, and scale is the key to understanding the systemic nature of the epochal choices that our forebears made. The automobile system that emerged from the cauldron of technological uncertainty at the end of the nineteenth century has become enormous. In many respects, it defies description. The alternative system would have been equally huge, performing many of the same functions and doing much of the same work as the internal combustion system.

To begin to appreciate the difficulty of situating the automobile within its systemic context, think about how we acquire and perceive information about the automobile. A picture of the vehicle itself, the artifact, inevitably stands out against whatever perceptual backdrop we create. It is the product shot in the television ad campaign. It is the brightly colored object on the gray concrete or black asphalt road. It is the discrete box painstakingly engineered by highly paid professionals to achieve a specific visual effect. In contrast, the road and other components of the automobile system form an endless and amorphous grid perceived only in bits and pieces or with the help of abstract maps. Perhaps most important, the vehicle is bought and sold in a very personal act. The rest of the automotive transportation infrastructure is external to the private moment of consumption. We are taxed or pay tolls for roads, we suffer mechanics, we pump gasoline, but we choose our cars. In every dimension, system function or mobility is concealed by or transformed into characteristics of the artifact. At every turn, the system frame dissolves, leaving only the vehicle in focus.

Contemporary studies of planning and transportation often recognize the transportation system as a collection of distinct but interacting components or subsystems.[2] Although we may not use the planners' language of intermodal splits, vehicle miles traveled, and fare recovery rates, every traveler can intuitively grasp the conceptual building blocks of transportation planning. The demand for transport service starts with the impulse to be mobile. For commerce or pleasure, we wish to transport ourselves (as passengers), our things (as freight), or our ideas (as information) from one place to another. At the most general level, transportation is a system in that it demands the coordinated actions of many different components. In the late nineteenth century, for instance, a traveler needed to know at least the extent and condition of roads, the availability of space for freight or passenger travel, and train schedules and fares. Moreover, each link in the transport process integrates both social and technical considerations. The options available at any point in time existed because of prior choices: should resources have been allocated, say, to build railways instead of roads or vice versa? Should steam vehicles have been permitted to travel on urban roads historically reserved as public space? Who decided which transport option to use? These choices had obvious implications for subsequent technological developments, but in no case were such answers driven purely by social or technical considerations. Options existed because of earlier decisions; past circumstances framed future choices. The physical hardware was fixed (i.e., determining) in the short run, but it was nonetheless the result of prior social decisions. Traditional scholarship on the history of the automobile has not employed a system perspective, yet the system view is necessary to understand how the interplay between temporarily fixed components of the transportation system and individual historical agents impelled the process of technological standardization toward the choice of internal combustion.

THE AUTOMOBILE AS MARVEL

To arrive at a new view of the automotive system, we must first understand the extent to which prior historical evaluations have downplayed the systemic character of the automobile. For much of this century, the historiography of the auto industry was largely the fabled story of individual entrepreneurism, managerial capitalism, consumer satisfaction, and increasing prosperity. The standard history of the American automobile went something like this: in the early 1890s, clever engineers and tinkerers, having seen the popularity of the safety bicycle and imagining a possible market for an advanced, motorized road vehicle, built hundreds of different motor vehicles, powered variously by electricity, gasoline, and steam. Although steam and electricity each showed early promise, gasoline was quickly recognized as the superior technology. By the early years of this century, steam and electric cars were already on the decline. In a triumph of technological creativity and ingenuity, American engineers and manufacturers, led by the inventor of mass production, Henry Ford, soon turned the automobile from a marvelous plaything for the rich into a mundane, and later mandatory, technological accouterment for the average American. By the outbreak of World War I, the automobile transport system had settled into a stable social and technological configuration: automobile equals gasoline power equals individual private transport. The market had spoken; the best technology had won. In 1914 American automobile manufacturers produced 568,000 vehicles. Approximately 99 percent of these were powered by internal combustion engines.[3]

During the ensuing decades this triumph became enshrined at home and throughout the rest of the world as an image of the internal combustion automobile literally propelling humanity forward. As one fawning account had it,

> the real story of the automobile is more wonderful than the fanciful
> tale of Aladdin's Lamp. It is more romantic than Romeo and Juliet.
> It is more important than anything else in the world, because it deals
> with the latest and by far the greatest phase of the art of trans-
> portation. Transportation has been the ladder upon which human-
> ity has climbed, rung by rung, from a condition of primitive savagery
> to the complex degree of civilization enjoyed by man in the 20th
> century.

The motor vehicle, according to this view, represented "the only improvement in road transportation since Moses, and the most important influence on civilization of all time."[4]

Moses might beg to differ, as might Gutenberg or Newcomen.
But while later scholars would moderate their rhetoric, the simple
story of pioneering grit, determination, and technological and entrepreneurial prowess, unfettered by state regulatory intervention,
remains popular and alluring. Walk through a contemporary automobile museum, and you will likely find the lineal progression from
the earliest to the most recent internal combustion–powered vehicles.
Electricity and steam, if mentioned or represented at all, will be tucked
into the corner along with other oddities.[5]

With respect to the choice of motive power for the automobile,
the preceding account assumes what it sets out to prove: that internal combustion was the best technology available. The technical, intrinsic characteristics of the gasoline engine were credited with being
responsible for its own inevitable triumph. *Automobile* and *gasoline*
have since been linked to each other by multiple and overlapping
cultural, political, and technological developments. Accordingly, the
expectations and service demands of alternatives to the automobile
are viewed through the lens of the intervening decades of the century. In this view, steam and electric vehicles suffered from a number of well-documented problems that rendered them ill-suited to
perform the service we now identify with the gasoline automobile.
Electrics were too slow, too heavy, and incapable, by virtue of their
battery-limited range, of car touring, the preferred activity of early
motorists. Steamers were slow to warm up, too complicated, too dangerous, and generally unmanageable for the average Sunday driver.
Automotive historian James J. Flink, writing in 1970, summarized the
orthodox view of the competition among steam, gasoline, and electric vehicles: "No one has yet presented a convincing argument that
the invariable association of the gasoline automobile with the cre-

ative automotive engineer-entrepreneur was due to anything other than the inherently superior technological feasibility of the internal-combustion engine over steam and electric power for the motorcar at that time."[6] In the best progressive tradition, the winning technology was the best because it won and won because it was the best.

SECOND THOUGHTS

Where were the alternatives before they became branded inferior technologies? By what criteria was internal combustion judged to be the best available technology? How did the characteristics of the gasoline engine become desirable? If gasoline was inherently superior, why were steam and electric vehicles still being produced into the 1920s, well after the dominance of the gasoline motor car was assured? The answers to these questions turn on who had the power to decide what the car should do and how it should do it. Internal combustion proved best at satisfying the demands of wealthy urban men, who wanted to tour quickly without being confined by the availability of charging stations. Internal combustion was also adaptable enough so that eventually its variants were able to satisfy the demands of a broader market of motorists.

Over the course of the twentieth century, Americans' perceptions of their cars changed. During the interwar years and immediately after World War II, the automobile was embraced without reservation. After many decades of unconstrained deepening and expansion, the automobile system finally began bumped up against social and environmental limits in the late 1950s. Air quality in Los Angeles and other major metropolitan areas bore witness to the hazards of unbridled automobility. Critics also addressed the direct health consequences of driving. Not since the earliest years of the century had public attention focused on the number of Americans injured and killed in motor vehicle accidents. By the middle of the 1960s, Ralph Nader had published *Unsafe at Any Speed,* a scathing critique of the neglect of safety in the design of American automobiles. A year after the book's publication, Congress approved the National Traffic and Motor Vehicle Safety Act, the first national legislation regulating construction and design practices in the automobile industry.[7]

Seizing on this crack in the cultural image of the automobile and reacting to the domestic gasoline crisis that followed the imposition of the OPEC embargo in the fall of 1973, historians of technology and transportation began to rethink the progressive, elegiac account of the emergence of the American automobile system. Several new directions proved ripe for investigation. Urban historians turned their attention to the relationship between the spread of the car and changes in urban structure. Within the larger urban setting, the automobile and its associated infrastructures (local roads, freeways, traffic lights, parking facilities, gas stations) became favorite subsystems for analysis. The revisionist position argued that Americans were promised more than they had received. Serviceable local transit systems were sacrificed on the altar of automobility. State and local issues, along with debates between urban and rural interests, were swept aside by increasing federal commitment to road building, culminating in the post–World War II decision to build the interstate highway system. Again, the automobile, in this view, was not the cause of suburbanization; rather, American tastes and values drew people to the suburbs, and the private passenger car was merely one institution that helped facilitate their move.[8]

These structural assessments of the role of the automobile in the growth of American cities examined some of the liabilities of car-based transportation but stopped short of considering the fundamental character of the automobile; its physical composition was never questioned.[9] Meanwhile, a second strand of scholarly discourse looked more closely at the history of the automobile system itself. Employing what sociologist Claude Fischer has called the billiard-ball model of socio-technical change, in which new technologies "roll in" from outside and "impact" society, these works looked at the social changes wrought by the coming of the automobile.[10] The most comprehensive and original work of this school was undoubtedly James Flink's *America Adopts the Automobile, 1895–1910* (1970), the first to characterize the minute interactions between the automobile and American culture that set the stage for the unprecedented growth of the automobile system. Flink addressed previously overlooked topics such as automobile clubs, regulatory and institutional responses to the spread of the automobile, and the rise of maintenance and repair shops. The weak point in his work and that of other billiard-ball analysts was their uncritical acceptance of the internal combustion–powered, pneumatic-tired, steering wheel–guided, privately owned passenger automobile as the single standard of automotive technology. Only in the past fifteen years have scholars attempted to flesh out the full range of automotive technologies and cultural variables involved in the larger transnational process of transport motorization. Today, the standard automobile and the traditional American story have begun to yield to the reality of a number of distinct motor vehicles, each solving different problems in different social settings.[11]

WHERE ARE THE ALTERNATIVES?

Scholars have studied the history of specific alternatives to internal combustion; but here, too, system-level issues have been overlooked in favor of discrete technologies. Nowhere is this trend more evident than in the writing on the history of the steam vehicle. For instance, in the portrait that emerges from Charles McLaughlin's 1965 study of the Stanley Steamer, the Stanley brothers are portrayed as master craftsmen wedded to nineteenth-century shop practice. The brothers were tinkerers, musical instrument makers who happened upon a uniquely efficient and lightweight boiler design in the late 1890s and then turned their considerable creative energies toward improving the technical elegance of their steam vehicles. They disdained marketing, labor specialization, and mass production; in short, they rejected growth, the sine qua non of corporate success, and opted to remain a small-scale vehicle manufacturer selling only as many cars as they could comfortably make in their Newton factory. The Stanleys, in this view, were antiquarians in their own time, throwbacks from the start.[12] Several popular studies in the early 1970s briefly considered the feasibility of reintroducing steam cars, but most scholars have been content to view the steam vehicle as a historical curiosity rather than a legitimate alternative to internal combustion.[13]

Studies of the electric alternative are more numerous and embrace a wider range of perspectives. Nevertheless, framing the rise and fall of the electric vehicle poses a methodological problem: the vehicle falls between several well-characterized technological systems and their respective literatures. There are different electric vehicles partially submerged in each of these different treatments. None is technically wrong; but to do justice to the history of the electric vehicle, it must be salvaged from its marginalization in each of these important bodies of scholarship and placed at the center of its own integrated account of social and technological change. In this sense, the following chapters are not merely stories about the electric vehicle;

they also demarcate an area around the technology to enable future scholars to distinguish it from the other host systems in which it has been concealed.

First, as an electrical device, the battery-powered car can be seen in the context of the expansion of the electrical network. Its fate can be linked to that of its supporting technologies: the lead-acid storage battery, the electric manufacturing industry, the prevailing network of charging stations, the rise of the electrical engineering profession, the cultural image of electricity, and the infamous "battle of the systems" between direct and alternating currents. This electric vehicle was little more than a minor detour, a trivial dead end, in the otherwise successful growth of the mighty electric industry. Works such as Thomas Hughes's standard history of Elmer Sperry, Ronald Kline's *Steinmetz,* and Harold Platt's *The Electric City* all put the electric vehicle at the margin of generally flourishing careers, institutions, and ventures. No two of these studies treat the electric industry the same way, yet all see the electric vehicle as a dead end.[14]

A second electric vehicle emerges from the study of changing patterns of urban transportation. Again, studies such as Kenneth Jackson's *Crabgrass Frontier,* Brian Cudahy's *Cash, Tokens, and Transfers,* Charles Cheape's *Moving the Masses,* and Mark Hirsh's *William C. Whitney: Modern Warwick* take differing views of the causes and consequences of electrification of the horsecar lines; but each dismisses the "trackless streetcar" as a failed spinoff of a basically promising technology.[15]

Third, whereas the literature on the steam vehicle has stressed the nostalgic appeal of this bygone technology, the historiography of the electric vehicle has been much less charitable. When the lens of history was turned on the electric alternative to internal combustion (for instance, by leading automotive historian John Rae), the critique was devastating. In a 1955 article on the Electric Vehicle Company's attempt to establish an automotive monopoly using electric taxicabs for urban service, Rae described the venture as a "parasitical growth" on the industry and argued that the "Electric Vehicle Company was founded on an error of judgment. It had put its money on the wrong horse—or, to be more accurate, the wrong horseless carriage." Other more general treatments also relegated the electric vehicle to the margins of the standard story of the automobile.[16]

More recently, the revival of interest in the future of the electric car has been accompanied by a modest outpouring of writing about its past. The artifact itself has been the subject of most of this work, but several scholars have tried to look beyond the boundaries of the vehicle and evaluate the relative importance of other, systemic factors.[17] Many of their essential discoveries and insights will be explored in the chapters that follow.

COMPETING SYSTEMS AND STANDARDS

Since the early 1970s, the historiography of the automobile has begun to emerge from the narrow confines of the standard story. But to take the next step—to look under the hood of the car into the black box of the automotive prime mover and then to connect the physical artifacts to the larger transportation system in which they were embedded—requires borrowing methods and perspectives from outside the mainstream of scholarship on the automobile. The fact that most studies of the history of the automobile have failed to admit that the choice of internal combustion was ever in doubt underscores the importance of looking more deeply at the choice of automotive prime mover and at the phenomenon of technological choice itself.

At the heart of the debate over the early history of the automobile are questions basic to any study of the history of technology: given the complex set of historical agents involved in an emergent social process, what directs, shapes, or dictates the outcome of a contest to establish or stabilize a technological artifact? In other words, what determines the result of a technological competition? Answers have tended to follow one of two paths: social or technological determinism.[18] Each explanation embraces a different range of specific views (for example, "soft" for social determinism and "hard" for technological determinism), but the question essentially turns upon the internal logic of technological change. Those who embrace variants of technological determinism argue that the outcome of technological competition is governed by the internal (or technical) characteristics of the artifacts in question, that there are such things as better and worse technologies, and that scholars can employ an impact assessment model—such as Fischer's billiard-ball model—to understand the relationship between technology and society. The social determinists, by contrast, seek to demonstrate the socially constructed nature of seemingly internal technological characteristics and argue against any unidirectional view in which technology influences society. In this view, technological closure—the process by which a technological standard is selected—is first and foremost a social process in which the relevant interests (i.e., public institutions, consumers, engineers, corporations) converge around and endorse a given technological option.[19] In practice, students of technology have recognized the importance of both perspectives. Technology has long been known to embody both social and technical components; and if explicit theories of the history of technology were slow to adapt to this reality, the beliefs and practices among historians of technology have admitted a role for both social and technical forces in the shaping of complex artifacts.[20]

Conspiracy theorists may be dissatisfied with this contextual synthesis. Indeed, in this case, as in a host of other instances of technological competition and standardization, both historical and contemporary, we want to *know* whether the best technology won or not. But simple explanations of complex processes are often wrong.

Again, the system lens is crucial to understanding how technologies compete. In the present case, "better" and "worse" were context-specific. The different engine designs competed in the real-world setting of the turn-of-the-century United States, not merely in disembodied form in an engineering laboratory. Already, therefore, we must move far enough away from the technical domain to know the systemic configuration of the competitors. To speak, for instance, of the electric car, we must recognize that the electric alternative encompassed a storage battery, an electric motor, gearing, tires, roads, charging apparatus, a primary source of electric power, private firms, and an array of other active and passive participants. It is important, in short, to know what is actually competing; the agents might be people, firms, ideologies, institutions, or governments. Even in the abstract and schematic world of theory, gasoline and electricity competed against each other as systems, not as stand-alone artifacts or equations on a chalkboard.

The following chapters describe these competing systems in increasing detail. Potential power sources, associated infrastructure, and their respective proponents are the actors or agents that were competing against each other to establish a standard automobile. Certain standards, such as driving on the right side of the street, were already established by social convention. Some technical standards emerged relatively painlessly from the early efforts of the nascent industry to self-standardize. Others, such as speed limits and access to various public and private spaces, were established by legislation after the technological configuration of the passenger automobile had been fixed. And behavioral standards, such as looking both ways before crossing the street, were learned the hard way as injuries and deaths from automotive mishaps began to mount.[21] Whereas most of these standards had relatively little impact on the emergence of a stable technological system, the choice of prime mover was crucial for the growth of the larger system because of the path dependent characteristics of the selection process. Could multiple standards have persisted? Was it inevitable that a single standard came to dominate the system? Perhaps separate spheres—for example, commercial versus private or touring versus urban—could have maintained separate technological prime movers.

In the case of the automobile, the dominance of internal combustion since the end of World War I has been nearly complete. Although the petroleum economy has never been as fixed and immutable as it appears, there is no denying the remarkable expansion of the private passenger–based transport system. From 1920 to the early 1960s, American social, political, and economic dependence on the automobile—what James Flink calls the "car culture"—crystallized into its baroque postwar formula: one car for every able-bodied person, ages sixteen to seventy, with average annual mileage rising from slightly less than 8,000 in the 1920s to just over 10,000 by 1970.[22] One may quibble with the elegiac account of the spread of the automobile; but in absolute terms and by any measure, the success of internal combustion has been astounding.

The automobile system also exemplifies long-term patterns of technological and environmental change. Technological systems enter new markets as novel and "clean," only to gradually become "dirty" as they expand in scale.[23] For example, the first automobile and the first commercial chlorofluorocarbons (CFCs) posed little environmental threat and may in fact have been more benign than the technologies they replaced. Moreover, the choice of the first replacement technology has nothing to do with its eventual health or environmental hazard. But after hundreds of millions of vehicles have been built or millions of tons of CFCs exhausted into the atmosphere, the environmental carrying capacity is severely tested. Eventually, if the expanding technological system is unable to continue to grow, new opportunities appear at the margins of the system, and the process begins again. Thus, it is no coincidence that doubts about the wisdom of continuing to expand the automobile system began to arise when they did, but not earlier. The problems of full-blown automobility were the consequence of and predicated on the success of full-blown automobility. Practically speaking, environmental transformation was an inevitable result of the motorization of the transportation system. In this respect, critics of internal combustion have underestimated the social and environmental consequences associated with other transportation alternatives. The costs of our dependence on gasoline are so evident—social isolation; environmental pollution; traffic jams; excessive commutes; road rage; limited mobility for children, elderly, and the disabled; and so on—that we are inclined to believe that the alternative must be better. In fact, we do not know what the specific liabilities of an electric-based road transportation system would have been, but we can assume that they would have eventually emerged as important constraints on system expansion.

AN EXAMPLE: THINKING ABOUT THE BATTERY

If you have read anything about the history of the electric vehicle, you may have come away with the idea that the electric vehicle failed in the battle with internal combustion because the lead-acid battery was unable to satisfy the range required for early motor vehicle applications. This straightforward line of determinist reasoning is indeed seductive, but it is only half right. It is correct inasmuch as the early storage battery posed a series of important challenges for automotive engineers. Rather than surmounting these challenges, however, the engineers side-stepped them with improvements to the internal combustion alternative. But blaming the battery does not provide a full explanation. If the outcome was governed by the energy density of refined petroleum, and since uranium has greater energy density than gasoline does, why don't we drive fission-powered vehicles? The answer, of course, is that we use energy in the system context. If you drive an electric vehicle and purchase your power from a utility that operates a nuclear power station, then you are driving a fission-powered vehicle. What if we reconstruct the historical experience compressed into the statement "the battery failed" from the system point of view? What if it was the battery-powered electric vehicle transportation system that failed in competition with the internal combustion system? Even this modest reframing leads to new and exciting questions: what did the electric vehicle system fail to do? For whom? In short, it was much more than the battery that determined the failure of the electric car. Accepting the limited, narrow explanation conceals the larger process of social, technological, and cultural adaptation that gave rise to what we know today as the automobile. Thus, the shorthand determinism explicit in the blame-the-battery explanation continues to cloud present debates about the future role of alternatives to internal combustion.

The automobile itself should not be considered a given. Rather than argue about the costs and benefits of the spread of the gasoline-powered automobile throughout society in the twentieth century, we must return to the beginnings of the automobile, when the heart of the machine was being contested. Society created the automobile; even the staunchest technological determinist must admit this fundamental truth. But what did it create the automobile to do? And how did the stable technological configuration that emerged satisfy social needs better than the available alternatives might have? Why, for instance, was it important that automobiles be capable of traveling long distances at high speed instead of simply providing dependable, usable, clean, local service? To answer these questions, we must move beyond the debate between the "good" auto and the "bad" auto and seek to understand the automobile as a material embodiment of the dynamic interaction of consumers and producers, private and public institutions, existing and potential technological capabilities, and prevailing ideas about gender, health, and the environment. The automobile came to symbolize a specific constellation of social objectives—speed, technological prowess, the experience and conquest of road-accessible nature. We are locked into this system now; but as we look forward to the second century of the automobile, we can envision it in a new and different social context. The automobile system was always flexible (in theory), but in practice there were limits to this flexibility. These limits were set not by the workings of the internal combustion engine per se but by the inertia of the technological system that grew up around it. Alternative systems were available but only at specific points in time. For a host of reasons, they were not chosen.

PART TWO

HISTORIES

2

WILLIAM C. WHITNEY, ALBERT A. POPE, RICHARD W. MEADE, AND THE ELECTRIC VEHICLE COMPANY

Semi-Public Electric Vehicle Transportation Service, 1897–1912

It is very evident to the average observer that the so-called horseless carriage is looming up as an important factor in the urban transportation problem. Animal power for the propulsion of street cars has been almost entirely superseded, in this country at least, by electro-mechanical power, and the horse seems to be threatened with further degradation by the substitution of like power for the propulsion of wagons and other vehicles in city streets. In this city [New York] a company has taken the matter up in earnest and is giving the public an electric cab and carriage service which for its reliability and comparative perfection is justly entitled to the support and admiration of all lovers of progress and enterprise. . . . the showing made in this instance, where a new industry is concerned, is indeed remarkable and reflects great credit upon the promoters for their foresight; upon the engineers for their skill in solving all the problems involved in the development of this industry; and upon the managers for the excellence of the service rendered to the public. There is every reason to believe that the electric vehicle industry is well established on a sure foundation and that it will grow rapidly, especially in the estimation of the public, without which support no enterprise of a semi-public nature could long exist.[1]

Editorial, *Electrical World*, August 1897

Electricity is the natural medium for the application of motive power. Its supply is unlimited. It is everywhere. It is to movement what the sun is to growth.[2]

Western Electrician, January 1898

The "electric cab and carriage service" described in the epigraph was inaugurated in New York City in March 1897 by Henry Morris and Pedro Salom, two Philadelphia-based engineers, with financial and logistical support from the Electric Storage Battery Company. The engineers, wary of entrusting the operation of their experimental electric vehicles to untrained and unsupervised drivers, had opted to establish a limited cab service instead of directly selling or leasing the vehicles. During the ensuing years, the company was reorganized and expanded several times. In 1899 the Electric Vehicle Company (EVC), successor to Morris and Salom's initial venture, merged with the Motor Carriage Division of the Pope Manufacturing Company, creating a massive holding company controlled by New York financier and transportation magnate William C. Whitney. For his part, Colonel Albert A. Pope had already amassed a sizable fortune selling Columbia brand bicycles, and Whitney and his partners were in the final stages of consolidating their ownership of all the major public utility systems in the city of New York. Together, their great ambition was to establish a national monopoly in the market for urban transportation service, combining electric omnibuses and taxis with existing rail-based transit systems. To this end, the EVC established regional operating companies offering electric cab service in major urban areas—Boston, Chicago, New York, Philadelphia, and Washington, D.C.

With Whitney's financial and political connections and Pope's management and production experience, the stage seemed set for economic success: the EVC would become a leading producer of motor vehicles, and its operating subsidiaries would use these vehicles to provide popular and affordable transport services in the growing cities of the turn-of-the-century United States. As an integrated provider of transportation service, the Whitney-Pope conglomerate would have been able to operate electric streetcars where economical, provide electric omnibus service on less traveled routes, and offer individual electric cabs where door-to-door service was required. In this vision, typical transit customers would have been free to choose the level of service they needed; in bad weather or when carrying heavy packages, an electric bus would take them the few blocks from the streetcar stop directly to their home for an extra dime. The stand-alone, motorized artifact that came to be called the automobile would be owned by this corporate entity, operated by its paid drivers, housed and maintained by the same experts who looked after the streetcars, and, most important, powered by electricity produced at the streetcar company's own generating station.

As we know, none of this came to pass. Whitney and his managers were unable to achieve the level of service integration initially envisioned. By the end of 1901 the organization was in disarray; the operating companies in Chicago and Boston had been shut down; and the larger automobile market had swung decisively towards private ownership, long-distance touring, and internal combustion. It is significant, however, that in spite of the evident failure of the integrated urban transit scheme, the early-generation cabs of the EVC continued to operate in New York for more than a decade; and promoters of internal combustion did not dare venture into the urban cab market until 1906.

Using company records, newspaper and journal accounts, and internal corporate communications, we can trace the development of the venture from the first demonstrations of Morris and Salom's Electrobat vehicle on the streets of Philadelphia in 1894 to the eventual demise of the EVC taxicab fleets in New York in 1912. Consigned to footnotes and sidebars in prior treatments, a full account of this far-flung enterprise has never been assembled. Beyond the financial machinations of its owners and the apparent liabilities of its electrical technology, the company produced many notable successes in systems management and operations. No single factor explains the collapse of this alternative transportation system. The company's electric technology was far from perfect, but it was not the sole cause of the EVC's demise. Neither were corrupt management and financial depredation responsible for the company's fate. Perhaps owners and managers should have done better; but in the end, the company was unable to create and sustain a working, integrated technological system capable of delivering affordable electric transportation service. Despite its extensive network of suppliers, employees, and consumers, the whole was less than the sum of the parts. Nonetheless, the history of the EVC reveals the contours of an inchoate transportation system that differs radically from the one we ultimately inherited and depend on. Had the EVC succeeded in establishing profitable operating companies in major urban areas, and had those companies attracted customers, suppliers, and infrastructure providers, the alternative, electrocentric, service-based system might have become the standard for local and regional transportation.

At its height the Electric Vehicle Company was both the largest vehicle manufacturer and the largest owner and operator of motor vehicles in the United States. With multiple assembly plants, operating companies in the half-dozen largest cities in the country, and sales agents from San Francisco to Mexico City to Paris, the EVC was also one of the first American motor vehicle makers to move away—however tentatively—from the small-scale production of custom-made vehicles that had dominated the emerging industry in the 1890s. Rather, its expansive, multidivisional corporate structure anticipated some of the innovations in corporate governance (Alfred Chandler's managerial revolution) that would spread through the rest of the automobile industry in the decade following the EVC's collapse. The company also pioneered a number of other important changes in the motor vehicle industry as a whole. In May 1897 one of its immediate predecessors, the Pope Manufacturing Company, became the first company to offer stock automobiles for sale to the public rather than producing them in direct response to orders. Another precursor, Morris and Salom's Electric Carriage and Wagon Company, was the first to sell motor transport service via its electric taxicabs rather than simply selling automobiles. This distinction—selling service versus selling automobiles—first established the principle that mechanized road vehicles could provide useful service beyond mere entertainment for the leisure class. Unfortunately, following its takeover by the Whitney-Philadelphia syndicate, the Electric Vehicle Company also became synonymous with trust building, stock jobbing, financial manipulation, and legal chicanery through the infamous Selden patent. The complex trust was beset by problems—from production delays and warehouse fires to shareholder suits and blistering public attacks. Although regional operating companies were established and perhaps 1,000 vehicles distributed to them, by the end of 1901 all but the New York, Philadelphia, and Washington branches were liquidated. The parent company was reduced to little more than a holding company for the contested Selden patent. An unfavorable legal decision and an economic downturn ultimately forced the EVC itself into default in December 1907, leaving only the reorganized component manufacturers and the aging cab fleet of the New York Transportation Company as testaments to the founders' dreams of electric cabs on every street corner in every major American city.[3]

Previous explanations for the failure of the expanded EVC enterprise have centered around claims that the electric vehicle technology was incapable of providing the required service—an argument for technological determinism, plain and simple. Eminent automotive historian John B. Rae has described the EVC as a "parasitical growth on the automobile industry . . . [whose] demise was regretted only by those unfortunate enough to hold its securities."[4] Subsequent scholars have tended to accept Rae's opinion, often generalizing from the fate of the specific venture to that of the electric vehicle itself. James Flink attributed the failure of the electric vehicle to the "superior technological feasibility of the internal combustion engine"; and even as recently as 1995, cultural anthropologist Michael Schiffer concluded that "it was not unsavory financial dealings that shook the cab companies . . . but failing batteries. . . . The cab operation was an extreme test of battery technology."[5] Deterministic arguments about the inherent liabilities of the electric battery have been augmented by claims that the managers and owners of the EVC engaged in a series of short-term financial manipulations that ultimately ran the cabs into the ground.

Neither technological determinism nor financial depredation can be entirely dismissed as contributions to the collapse of the EVC, yet a careful examination of previously unavailable documents suggests that other factors also played an important part in the final outcome. Although some issues, such as labor relations, were already visible in the early days of the EVC, others (for example, issues relating to vehicle design and production and to fleet operation and maintenance) came to light only after the creation of the expanded holding company. Looking beyond these problems, we can see that the story of the EVC is also a tale about the creative and determined managers, engineers, and drivers who succeeded in making the electric cabs go.

PHILADELPHIA, 1894–1896

Residents of Philadelphia first glimpsed a horseless carriage late in the summer of 1894, and energetic engineers Pedro Salom and Henry Morris spent that fall testing prototype electric carriages on the city's cobbled streets. Their vehicles, called Electrobats, were literally horseless carriages: wooden-wheeled wagons to which they affixed masses of lead-plate batteries and electric motors. Salom designed the battery system, having spent much of the preceding decade developing secondary batteries for streetcars for the Julien and Consolidated Companies, while his partner Morris adapted the equipment to the carriage, acquiring components from various existing producers.[6] Their work was supported by the Electric Storage Battery Company (ESB), the leading American manufacturer of Chloride-Manchester lead-acid batteries. By late 1894 the ESB had consolidated its hold on the domestic battery market through acquisitions of

competing firms and their patents; the domestic market for stationary batteries was finally beginning to grow, and the company anticipated expansion into new markets.[7]

The first Electrobat weighed more than 4,200 pounds, including
1,600 pounds for the batteries. Electrobat II used a lightweight
piano-box carriage design in which the batteries weighed only 160
pounds, approximately 10 percent of the total curb weight of 1,650
pounds. During the following year Morris and Salom tested and expanded the operating envelope of the two Electrobats, learning the
vehicles' effective power and range from experience. By the spring
of 1896, Salom estimated that the vehicles had logged several hundred miles.[8]

Morris and Salom were not the first Americans to construct a
four-wheeled, electrically powered horseless carriage, but they were
the first engineers to attempt to incorporate electric carriages into the
transportation system of late nineteenth-century American cities.[9] To
publicize their innovations, they entered Electrobat II in the famous
Chicago Times-Herald motor vehicle road race held in Chicago on November 28, 1895. An early winter storm left the roads treacherous
and heavy with snow, limiting the vehicle to a shortened demonstration run on race day. Having drawn down the battery during their
drive to the starting line at Jackson Park, Morris and Salom were concerned about the vehicle's range and, after two short stops to replace
spent batteries with fresh ones, decided to turn back after covering
only eleven miles of the fifty-three-mile course.[10] Despite this apparent setback, they returned to Philadelphia flushed with enthusiasm.
They had failed to win the $2,000 grand prize, but the judges had
awarded them a gold medal for excellence in design; and Morris and
Salom knew that their horseless carriage would work well under normal road conditions, regardless of its difficulties on Chicago's win-
try north shore.

Encouraged by their showing at the *Times-Herald* race, the two
men spent the following months back in Philadelphia envisioning
possible applications for their horseless carriages. Neither of the
Electrobats proved capable of the kind of long-range travel (more than
thirty miles on a single charge) required to compete successfully in
the *Times-Herald* race or in others scheduled afterward. Moreover, the
electric carriage, although easy to operate if well maintained and fully
charged, was a complicated piece of electrical technology. Some enthusiasts looked forward to the day when individually owned and

operated motor vehicles would replace all horse-drawn wagons and carriages, but Morris and Salom were more cautious. They did not believe, for instance, that untrained drivers were capable of servicing and maintaining an electric carriage. In a lecture at the Franklin Institute on April 4, 1896, Salom set forth their views:

> We do not consider it practicable, at the present time, to send out such vehicles broadcast over the country before any proper arrangements have been made for their intelligent care and maintenance. We think that the proper plan for their introduction is to construct a sufficient number of vehicles of one kind to warrant the building of a charging station, where the batteries can be charged and the vehicles kept when not in use. . . . A building in a suitable central locality can be purchased or leased, and the vehicles can then be operated either on a lease or rental plan, very similar to the manner in which a livery stable is at present conducted, thus assuring to the owners or lessees a complete and reliable service, such as they could not secure at first by attempting to operate the vehicles themselves.[11]

Whereas other electric vehicle designers like Andrew Riker focused on building vehicles to compete in increasingly popular motor vehicle races, Morris and Salom intended to use their electric carriages neither for long-distance rural touring nor for short-track racing but to replace horses in urban parks and boulevards.

HARTFORD, 1895–1897

While Morris and Salom were busily making and testing their Electrobats in Philadelphia, several dozen other entrepreneurs were also hard at work building horseless vehicles. Following the widely publicized *Times-Herald* competition, their number blossomed into the hundreds. From small metalworking shops in New England and the midwest to the design studios of established bicycle and carriage manufacturers, creative energy was directed toward solving the problems of motorized road transport. Even so, in early 1895 there were probably no more than fifty people in the country who could claim to have successfully built and operated a horseless carriage.

Hiram Percy Maxim, a young engineering graduate of the Massachusetts Institute of Technology, was one of the select few. As he recalled in *Horseless Carriage Days,* his memoir of the early history of the American automobile, Maxim had already spent many spare hours trying to fit a homemade gasoline engine to a second-hand tricycle. In June 1895 he was hired as chief engineer of the newly constituted Motor Vehicle Department of the Pope Manufacturing Company in Hartford, Connecticut. Arriving in July, he reported to Hayden Eames, a gifted and energetic former naval officer. Maxim quickly set about the task of building a four-wheel prototype for his new employers.[12] The resulting vehicle, called the Crawford, was a lightweight, piano-box carriage powered by the same engine he had used in his experimental tricycle. He thought it was a good first effort, but the Crawford did not impress the Pope managers. It was too noisy, it rattled as if it were about to fall apart, and it was so plain in appearance that general manager George Day could not imagine how to sell it or who would buy it.

Maxim was sent to Chicago as an official for the *Times-Herald* race. Each vehicle entered in the race was assigned its own umpire to verify that the operators did not violate any rules, and Maxim drew the Morris and Salom Electrobat. Although already attached to the idea of internal combustion, Maxim witnessed firsthand a functioning electric vehicle just as he was facing enormous barriers to the effective application of internal combustion. His memoir records his feeling at the time: "I saw that building a sweet-running little gasoline carriage was going to be about a hundred times bigger job than I had expected. . . . Only courageous men well equipped with tools, knowledge and spare parts, and indifferent to dirt, grease, smoke and noise, could consider going anywhere in any motor vehicle thus far produced."[13] It was time to go back to the drawing board.

After returning to Hartford, Maxim began work on a completely new vehicle: the electric Mark I carriage. To validate the idea of a production motor vehicle (not to mention save his job), he constructed the Mark I with the intent of reassuring the company's senior managers that they were not making a mistake by committing themselves to the motor vehicle market. Catering to their tastes required Maxim to pay attention to the aesthetics of the carriage and to its general ease of operation. The Mark I Phaeton, as it was called, first rolled through the streets of Hartford in April 1896. Although the fine-grain finish of the carriage body did little to enhance its operating characteristics, all who rode in it were impressed. The carriage handled smartly, stopped and started with minimal effort, and, best of all, emitted no noxious fumes or noises. After careful demonstrations for the various officers of the company, Maxim made plans to produce a small batch for sale to the public in the spring of 1897. The ten vehicles, bearing the designation Mark III, were unveiled in a public ceremony in downtown Hartford on May 13, 1897.[14]

Privately, Maxim worried that his Mark IIIs were not yet ready to leave the factory testing ground. Even under his close supervision, the experimental vehicles had suffered abuse from untrained drivers, and Maxim struggled with "the knowledge that ten of my precious darlings were out on the road in the hands of ordinary human beings, who might not demonstrate them intelligently."[15]

ELECTRIC CARRIAGE AND WAGON COMPANY, NEW YORK, 1896–1897

By 1896 Morris and Salom were ready to take the first steps toward putting their horseless vehicle scheme into practice. The Electric Carriage and Wagon Company, with nominal capital of $10,000, was incorporated in January of that year. Among the founders was George Herbert Condict, a leading battery engineer who later played a crucial role in the expansion of the New York electric vehicle system.[16] The following year, the first experimental vehicles were rolling through the streets of New York. Soon, however, Morris and Salom relinquished ownership and managerial control to Isaac Rice and W. W. Gibbs, senior managers at the Electric Storage Battery Company. As Michael Schiffer has observed, by 1897 the Electric Carriage and Wagon Company "was no longer recognizably distinct from the ESB."[17]

In mid-March 1897 the preliminary tests were complete, and the electric cab system was ready for business. On March 27 the Electric Carriage and Wagon Company began offering electric cab service in and around Manhattan, charging rates equivalent to those of the horse-drawn cabs: one dollar for the first two miles and fifty cents for each additional mile for up to two people.[18] The vehicles themselves (built in Camden, New Jersey, by the Charles S. Caffrey Company) combined aspects of the standard horse-drawn cabriolet with the narrow pneumatic tires, spoked wheels, and massive construction of Morris and Salom's first vehicle (figure 2.1). Two 1.5 horsepower Lundell motors provided by the Interior Conduit and Insulation Company in New York City were capable of propelling the vehicles at an average speed of eight miles per hour, and the battery consisted of approximately eight hundred pounds of standard Chloride-Manchester

FIGURE 2.1. First-generation electric hansom cab, Electric Carriage and Wagon Company, New York City, 1897. *Source: "Electric Motor Cab Service in New York City—I,"* Electrical World *30 (August 14, 1897): 184.*

plates from the ESB.[19] An important design element of Electrobat II was also incorporated: the battery was composed of separate trays to facilitate rapid exchange. Subsequent generations of electric cabs would come to depend upon the capacity for easy battery exchange.

How did the cab station operate in practice? Although the vehicles were called cabs, most did not wait for customers at traditional hack stands. Rather, the vehicles were usually dispatched on call to hotels, clubs, and private residences (figure 2.2). Recognizing that the vehicles were largely untested, the dispatcher was instructed to be selective; "the impossible or impracticable was not to be attempted."[20] In other words, the vehicles were not to be overextended. A perma-

FIGURE 2.2. First-generation electric vehicle "on call"; an electric brougham and an electric hansom in front of the Casino Café, New York City, c. 1897. *Source: Collection of the Museum of the City of New York. Reprinted with permission.*

nent staff of six—a station superintendent, a starter, two battery specialists, a machinist, and a washer—supported twelve drivers with approximately ten vehicles in service on any given day. The total fleet consisted of thirteen vehicles: twelve for hire and one for service. The station was equipped to maintain twenty sets of batteries, which were removed from the cabs and wheeled to charging benches with the aid of a hand-operated transfer truck. Preliminary announcements promised that "competent engineers" would be hired in place of typical horse-cab drivers, and Pedro Salom himself reported that the company hoped to "improve the personnel of the service by having educated drivers . . . [with] more or less training in practical electrical engineering." But management quickly discovered that the "hackman's business is a profession. They have to know all about the city and how to go everywhere, and it is not every man you pick up who will meet the requirements." Running the vehicles, by comparison, was "a simple thing." The company quickly abandoned its initial strategy and hired former hack-cab drivers. The uniformed recruits were able to "take hold of the new vehicles very readily, appreciating their advantages, and . . . [soon became] very expert in handling them." According to Salom, the company encouraged responsible use of the cabs by insisting that the wagons could not be pulled home by a horse: "when the driver has to push his machine home, he does not often get left."[21] In the absence of more detailed operating accounts, it is impossible to know how strictly this rule was enforced. A poorly adjusted brake shoe or hot wheel bearing could result in excess energy usage and ultimately lead to battery failure, but we have no sure way of knowing the relative contribution of nonbattery mechanical problems to the overall reliability of these first-generation cabs.[22]

Operating data on the early fortunes of the electric cab fleet are scarce at best. The Electric Carriage and Wagon Company was not subject to any legal reporting requirements, so all we know is what the company chose to divulge. In the July 1897 issue of *Horseless Age,* Morris and Salom reported preliminary figures on their first eighteen weeks of vehicle service. With this report as a starting point, what can we say about the profitability of the electric service? From an operational perspective the cabs appear to have provided reasonable service, covering a total of 14,459 miles—more than 1,200 miles per cab. Accepting that ten vehicles were in use on any given day and counting 130 days between March 27 and August 3 (the station was reportedly open every day), the cabs averaged about eleven miles per day per vehicle and claimed to have carried 4,765 passengers. From subsequent records of the New York cab fleet in 1899 and the Atlantic City fleet in 1900, we can estimate that approximately two-thirds of total mileage was "paid" (about one-third were "dead" miles run to pick up a fare or return from a dropoff).[23] Using these data, we see that the electric cabs generated revenues of approximately forty cents per mile, probably enough to produce a small operating surplus.[24] These speculative assumptions and calculations are required because the company chose to remain silent about the revenue side of its operation. In reporting its accomplishments, the company may have been trying to advertise its successful entry into the transportation service industry, just as McDonald's restaurant signs once touted how many billions of hamburgers had been served. Revenues and eventually profits were likely expected to increase as the electric fleet expanded and its managers learned to improve productivity. Then, as now, young companies often did not expect or achieve profits during the startup phase of operations. Nonetheless, Morris and Salom achieved their primary objectives. The electric cabs established an alternative to horse-drawn passenger transport service and demonstrated enough potential for profitability to encourage ESB president Isaac Rice to enlarge the fleet.

FIRST EXPANSION: THE ELECTRIC VEHICLE COMPANY, 1897–1899

On September 27, 1897, the six-month anniversary of the inauguration of electric cab service, Rice incorporated the Electric Vehicle Company in the state of New Jersey.[25] As reported in *Horseless Age,* he planned to dramatically expand the number of cabs in service, promising to put one hundred more vehicles on the street as soon as possible. In the brief history of the American automobile up to that point, no one had placed such a large single vehicle order. Indeed, to place the magnitude of the endeavor in context, the entire national production of automobiles for sale in 1897 was probably not more than fifty vehicles. The largest manufacturer of gasoline vehicles was former bicycle producer Alexander Winton, who is recorded as having made and sold four cars that year, while the Motor Vehicle Department of the Pope Manufacturing Company produced several dozen Mark III electric Phaetons.[26] Before the EVC, automobile production was a skilled craft practiced vehicle by vehicle at the margins of existing lines of business. Often the first (and sometimes only) car produced was added to the stable of the shop owner. The proposed expansion of the cab fleet thus posed a considerable logistical challenge for the nascent industry.[27]

A decade later, Henry Ford took advantage of volume sales to initiate the mass production revolution, relentlessly forcing down unit costs as he strove for greater efficiency of production. But where Ford saw potential strategic advantage, the EVC saw only logistical problems: how could the company coordinate batch production among a farflung network of suppliers? Details about the first efforts to increase production at EVC are scant, but we know from Maxim's account of the design and production of the Mark III Phaetons in the same year that batch production essentially consisted of Maxim traveling down the Connecticut River to William Hooker Atwood's New Haven carriage shop and up river the next day to the Eddy Electric Company, maker of electric motors and controllers. Atwood's New

Haven Carriage Company was eventually folded into the EVC's holdings, but at first the only economies of scale or scope were in the mind of Maxim, the system integrator.[28] Within several years, the electrical supply industry became aware of the electric vehicle opportunity and marketed special lines of motors, converters, batteries, and charging outfits for electric vehicles. But in late 1897 electric vehicle manufacturing was far from an established line of business.

Further evidence of the prevailing inefficiency of production appears in later records of EVC engineer William Crane, who was responsible for the redesign of the third generation of electric cabs ordered in the winter of 1899. Even a year and a half after the EVC first tried to increase its scale of production, purchase orders and executed contracts with carriage makers show that several different shops were under contract and that no single maker received an order for more than one hundred vehicle bodies. Moreover, within a given order, there was never more than fifty of any given type of vehicle ordered at any one time.[29] Although the EVC was the first mass purchaser of automobile components, it never managed to become a mass producer of a standard automotive product. The net results were evident. On the one hand, the company paid too much for its capital goods, raising the overall cost of producing electric cab service; on the other hand, there was no such thing as a standard electric cab or a standard storage battery.[30] With various batches of different types of carriages roaming the streets, the idea of interchangeable parts so often associated with the American system of manufacture remained a dream.

Given the problems associated with scaling up and standardizing vehicle production, the task of systems integration was pushed down the manufacturing hierarchy. Because the EVC production process did not turn out a standard vehicle, the job of standardizing service fell to the new central station where, in a converted skating rink at 1684 Broadway, engineers, superintendents, battery technicians, cleaners, and drivers organized to mass-produce transportation service. The chief engineer responsible for designing this complex facility was George Herbert Condict. He had been involved with the electric cab venture from the outset but was not appointed to plan and construct the enlarged central station until late 1897. Condict brought crucial expertise to the EVC operation: several years earlier he had supervised the battery streetcars that ran on Fourth and Madison avenues in New York; and he had also been responsible for designing the equipment for the Englewood and Chicago electric storage battery streetcar line. In both instances, Condict had designed the plants to facilitate automatic, mechanical manipulation and exchange of heavy sets of batteries. Applying the solutions from battery streetcar operation to the EVC revealed further complications. Specifically, a free-wheeled cab, unlike a streetcar, was prone to shift up and down and side to side as force was applied to it. In response, Condict designed the battery exchange bays so that the vehicle was held in place by hydraulic positioning shoes that centered it. The cab was then lifted off its springs and the battery tray hooked to another hydraulic piston that pulled the entire battery set, weighing nearly 1,300 pounds, onto the adjacent conveyor table. An overhead crane plucked it from the table and deposited it in the charging room, while the vehicle received a freshly charged battery and was back on its way. Although simple in concept, the operation was enormously capital-intensive, demanded extreme precision, and was the opposite of practice in the thinly capitalized, horse-based transportation industries that dominated the urban transportation market. No fewer than forty subcontractors were needed to complete the station; and one observer, watching batteries being shuttled around the station, noted that "the manipulation resembles more than anything else the handling of steel billets in the reheating furnaces of rolling mills."[31] In short, Condict's central station was a marvel of modern mechanical engineering.[32] For the first time, industrial practice was brought to bear on the age-old problem of transportation over city streets.

The organization and construction of the new central station was only one of a series of technological changes implemented by the EVC. Taken together, these adaptations underscore the systemic nature of the challenges that the company encountered. Second- and third-generation cabs underwent design changes suggested by day-to-day problems and complaints: the tires were widened, the battery compartment was repositioned, the passenger cabin was enlarged, and the brakes were strengthened (figure 2.3). Whereas the prototypes used three-inch-wide pneumatic tires with different-sized wheels on the front and rear axles, subsequent versions used wider five-inch pneumatics mounted on solid disc wheels that were identical in front and rear. These changes were necessitated by the frequent and rapid failure of the narrower tires under the relatively severe road conditions of New York City. During Morris and Salom's quiet cruises in Philadelphia, thinner tires were not an issue; but as the vehicles were

NEW YORK ELECTRIC VEHICLE

TRANSPORTATION COMPANY,

1684 BROADWAY.

TELEPHONE, 381 COLUMBUS.

FIGURE 2.3. Later-generation electric hansom cab, as shown on brochure for New York Electric Vehicle Transportation Company, September 1, 1900. *Source: William F. D. Crane Papers (MG 1092). New Jersey Historical Society, Newark, N.J. Reprinted with permission.*

passed to hired hands, forced to shoulder heavier loads, and used in situations in which faster service increased revenue, narrow tires and spoke wheels soon proved unworkable. Writing in October 1898 to George Condict, engineer William Crane observed: "There are twenty cabs on the street and there appears to be business for several times that number, as we have calls every day that we have to refuse. The great difficulty at the present time is in the puncturing of tires. We have eight rubber concerns laboring with us on this problem and hope in the near future to get something that will stand the work."[33] Thicker tires, built especially for the EVC by the Diamond Tire Company, were adopted only after careful experimentation with other models, including solid rubber tires and "comfort" tires (solid tires with surface air pockets that provided some cushioning while avoiding the risk of puncture associated with the pneumatics).[34] As was the case in the overall design of the central station facility, components of the technological system were reworked in response to changes in the social context in which the vehicle was expected to perform. In the switch from owner-operated test vehicle to employee-operated commercial vehicle, the tire became a bottleneck to the success of the electric vehicle fleet. Redesigning the tire was essential for the venture to stand a chance of success.[35]

The battery compartment was redesigned in later versions of the electric cabs. Blueprints and engineering diagrams for the third-generation vehicles (dating from early February to mid-April 1899) show that the battery storage compartment was slightly expanded and reconfigured, perhaps to allow larger batteries to be carried but also to better ventilate acidic fumes and isolate corrosive sulfuric acid from mechanical components and wiring harnesses. Technical communications between the EVC and one of its carriage builders discussed the "special provisions . . . made for the running of wires and brake connections in compartments tight against the fumes and moisture of the battery."[36] This adjustment responded to the observation that "one of the greatest sources of trouble in the cabs is the slopping of acid on the working parts. In many cases it has been discovered that the acid was very much higher in the jars than necessary."[37] The relationship between battery storage and the mechanical and electrical components (like other problems addressed in the second- and third-generation cabs) could not have been easily foreseen in the design process. It was only after some months and many miles on the streets that the liabilities of the initial vehicle system became apparent.

Significantly, the available technical records and reports from this period do not contain extensive reports of vehicle failure as a result of insufficient battery capacity. Several explanations are possible; for example, the file with these complaints may not have survived to the present. Alternately, perhaps there were no battery complaints because the drivers, station crew, and engineers did not bother to complain about things they could not control. They may have simply accepted the limited range of the electric cabs as one accepted the range of a horse, the waning of the moon, or one's full-grown height. If it could not be changed, why waste time and energy complaining. To be sure, Salom admitted that the battery was a serious problem.[38] *Electrical World* opined that, in spite of recent improvements, the storage battery "will spatter, fume, give out on the road, leak, buckle, disintegrate, corrode, short-circuit and do many other undesirable things under the severe conditions of automobile work."[39] And there were occasional reports of cabs running out of power while on duty and having to be towed in. Perhaps at this early stage, however, these were exceptional and expected events, barely worthy of remark, let alone concern. Moreover, as Salom suggested, such incidents may have been interpreted as a sign of operator error or carelessness rather than a symptom of a failing technology. Maxim, for one, observed that new drivers quickly mastered the basics of operating a motor vehicle. Problems that were first attributed to the poor design of the machine were soon internalized so that subsequent incidents were blamed on the operator rather than the technology.[40] This interpretation makes no implied judgment about whether battery range was either adequate or not, nor does it deny that the lead-acid battery was extensively criticized. But batteries did tend to perform best during their first six months of service life; and relative to the range of competing horse-drawn and gasoline vehicles, the electric vehicle of 1898 did not appear as inadequate as it does in retrospect. Criticism tended to come not from people who had actually used storage batteries to operate motor vehicles but from outside observers.[41] Company files record no internal defections from the electric technology or even any retrospective misgivings about the choice of technology. In fact, those employees who were closest to the machines continued to believe in their serviceability long after the failure of the venture was apparent. As late as 1904, for instance, EVC sales manager Frank Armstrong was writing to senior management complaining about the company's lack of support for its electric vehicle sales and service campaigns.[42] Although criticism was enthusiastically leveled at the storage battery in the months and years that followed, none of these attacks came from EVC engineers, station crew members, or drivers. Even when New York Transportation Company president Richard W. Meade finally pulled the plug on the electric cabs in 1912, he did not do so on account of battery failure but in response to the improving financial opportunity for operating internal combustion cabs.

While the engineers and supervisors were weighing these and other technical issues, the drivers and their electric cabs were out on the streets of New York carrying passengers. Periodic accounts suggest that New Yorkers welcomed the electric vehicles and patronized them liberally. The appearance of the vehicles prompted considerable debate. One society writer, the eponymous Cholly Knickerbocker, claimed to have been "the first representative of dudedom who has ever ridden in a horseless carriage in the garish glare of day." The author compared riding in the cab to a dream in which "I walked down Fifth Avenue in my pajamas in the full tide of the afternoon promenade. . . . I had something of the same feeling as I sat there and felt myself pushed forward into the very face of grinning, staring and sometimes jeering New York." Although he gradually came to accept that he did not need "the protection of a horse in front," the writer nonetheless felt a "sense of incompleteness" from riding in the horseless carriage and concluded that something like the horse was needed to protect the privacy of the occupants: "as the vehicle is now constructed there is all together too much publicity about it."[43] In short, although this account stresses the novelty of the horseless carriage, perhaps more striking is the fundamental similarity between the electric carriage and the hansom cab that it hoped to replace. The electric carriages did not alter the distances traveled, the purposes for seeking a cab, the costs associated with hiring private transport, or any other behavioral aspect of the transportation process. The only apparent difference was that the builders had not bothered to adapt the carriage design to compensate for the absence of the horse and the consequent loss of privacy. Another editorialist was dismayed by the excessive conservatism evident in the design of the vehicles and pleaded "for some style of electromobile that will proclaim what the thing is and not let it go around forlornly as if it were an ordinary vehicle seeking the departed spirit of its horse."[44] But aesthetic concerns did not hinder enthusiasm for the new machines.

Demand for cab service regularly exceeded available supply. An

1898 article in *Electrical Engineer* reported on the "rapid tendency towards the adoption of automobiles ... shown ... in the fact that it was the correct thing to go to Mrs. Astor's ball this week in an electric carriage."[45] That December, shortly after William Crane had reported that demand for vehicle service could not be met because of problems with the tires, an early snowstorm struck New York City. In the era before snowplows, even a modest storm could snarl traffic for days; but because merchants and homeowners shoveled the sidewalks, the maneuverable electric cabs took to the footpaths and shouldered through the storm, handling calls that were forwarded from the horse-cab companies.[46] Attributing the success of the vehicles in part to the wide pneumatic tires that rode over the snow instead of cutting into it, reports claimed that the storm had "demonstrate[d] the superiority of the electric cabs" and that "the electric motor cabs were found to be more available than any other form of transportation during the heaviest part of the storm, and they did a large business in the hotel and theater district."[47] About the same time, Fred Vieweg, the EVC's general manager, was interviewed by *Horseless Age* about the state of the electric cab operation: "We are now renting cabs by the month, and number among our patrons ladies who maintain their own carriages, but prefer to use our cabs at night. Most of our drivers have a regular list of patrons. We pay our drivers $2 a day, and, in addition, those who are polite and accommodating pick up quite a tidy sum besides."[48] An editorial in January 1899 confirmed Vieweg's report: "it is difficult for love or money to engage one [cab] for the evening later than 10 o'clock in the morning," and an account of a meeting of the New York Electrical Society in February 1899 claimed that orders for the cabs were frequently "three deep between nine in the evening and twelve o'clock." That month, during another fierce winter storm, the members of the Electrical Society convened at the EVC central station. Although chief engineer Condict was trapped in Philadelphia by the snow, the visitors who braved the storm "saw the whole rolling stock of the company in active operation, literally coining money."[49] A handful of complaints about "incompetent and reckless" drivers was accompanied by lawsuits, and it is clear that the company was experiencing some growing pains; but on the whole the public seemed pleased with the new service and the company that provided it (figure 2.4).[50]

Given the total number of miles that must have been logged during the first two years of fleet operation, the documents that survive

FIGURE 2.4. Later-generation electric cab driving through snow-covered city street, New York City, c. 1899. *Source: Collection of the Museum of the City of New York. Reprinted with permission.*

offer only slivers of insight into the daily routines of the Electric Vehicle Company. A limited number of operating records have been preserved. Summarized in figure 2.5, they provide a brief snapshot of the electric cab fleet in late January 1899. Approximately forty-five cabs were in regular fleet service, and perhaps forty more were rented or leased to patrons under long-term service contracts.[51] For the

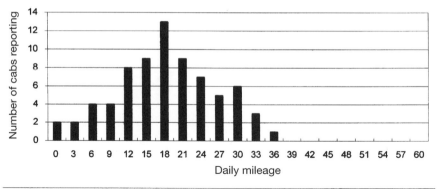

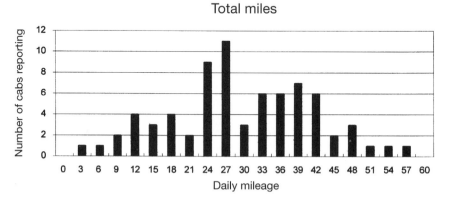

FIGURE 2.5. Summary of daily cab operations, Electric Vehicle Company, New York City, January 1899. *Source: William F. D. Crane Papers, New Jersey Historical Society, Newark, box 2, folder 7.*

vehicles in the daily cab fleet, average mileage was just over twenty-eight miles per day gross and seventeen miles per day paid. A handful of vehicles was apparently dispatched for single trips (that is, they had low total mileage, and dead and paid mileage were nearly equal); others covered well in excess of the average daily mileage. At least two traveled more than fifty miles in a single day. Translating this data into revenue, the fleet generated approximately $11.25 per cab per day. As before, the question of profitability is not easily answered; but comparing the fleet operation in January 1899 to that reported in August 1897, we see that average mileage and therefore average revenue had increased considerably.

Other operational issues, however, must also be considered. Reports indicate that relations between drivers and management were strained at best. Although the cabs might have been profitable in theory, there was no mechanism through which management could compel drivers to truthfully report their earnings. Maxim recalled that one of the operators confided "with an evil leer . . . [that] the cab drivers could 'beat' any plan the company devised to insure receiving all the money the drivers collected."[52] General manager Fred Vieweg tried to introduce profit sharing to link pay with amount of service provided. By limiting a driver's base salary and linking compensation to total revenue, the driver was ostensibly encouraged to be more forthright in reporting fares; but as Maxim's informant observed, there was no way to force the drivers to comply.[53] The driver could always "knock down" the fare that he reported on his daily service log and pocket the difference as extra tip. Obviously, management tried to discourage this practice, claiming that knocking down and overcharging went hand in hand. And in at least one instance, the company went so far as to prosecute one of its drivers for taking one dollar out of a paid fare.[54] The taximeter—a device for measuring distance and time intended to prevent passengers from being overcharged—was also considered. Internal communications of the EVC reveal that management had contracted with an engineer named James Keyes in 1898–99 for the development of a "taxameter" but decided in August 1899 not to pursue the matter and to continue to rely on manual reporting mechanisms.[55] The mechanical fare meter was not introduced in the United States until 1907.[56] Thus, it is clear that during the first few years of motor cab service, the profitability of the company turned on more than simply building fleet mileage.

The unfamiliarity of electric transport service also influenced the question of profitability. In 1897 electric vehicle service was still a new transportation alternative for turn-of-the-century New Yorkers. To the extent that some patrons might have been willing to deliberately seek out electric cabs instead of settling for a typical horse-drawn hack, the electrics probably benefited from an inflated market demand for their services and a general willingness to accept the growing pains associated with the expansion of the fleet. By 1899, for instance, managers at leading hotels had begun to request automobile service for their patrons; so it is possible that as a result of the novelty value attached to the new mode of transport, actual revenues were higher than vehicle mileage alone would predict.[57]

Any discussion of profits must also consider who stood to gain from the success of the Electric Vehicle Company. The ESB was, in part, a holding company for EVC shares. Because the major stockholders of the EVC were the owners and directors of the ESB, producing operating profits from the fleet may have been less important than establishing a stable, captive market to justify the ESB's continued expansion.[58] As long as the owners of the EVC were able to expand the cab fleet and avoid losing too much money, they were guaranteed a reasonable return. Financial records from the time suggest that they did just that. As figure 2.6 shows, ESB business expanded dramatically from 1898 through 1900, the years during which the EVC venture was on the rise. This kind of market subvention, although questionable from a short-run economic perspective, was neither illegal nor improper. Underwriting modest operating losses involved in the creation of a new business unit, however, is different from the improper and damaging speculative frenzy that soon followed.

Why dwell on the question of profitability? First, the issue of who knew what and when has been debated by observers of the Electric Vehicle Company. According to the editors of *Horseless Age,* the company knew it had embarked on a money-losing venture almost from the start, and subsequent investors merely tried to expand the business and conceal its losses long enough to unload their own shares. Modern scholars have also disagreed about when the principals of the EVC learned the real financial condition of the company. Second, the question of profitability bears upon the subsequent evolution of the company. If it was indeed a profit-making enterprise in early spring 1899, then there might have been good business reasons behind the move to expand the operation. But if the company al-

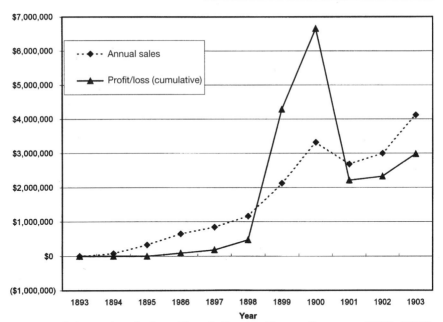

FIGURE 2.6. Sales and profit/loss, Electric Storage Battery Company, 1893–1903, in the period leading up to and including the expansion of the EVC's electric cab venture. *Source: Andrew Irvine, "The Promotion and First Twenty-two Years' History of a Corporation in the Electrical Manufacturing Industry" (M.A. thesis, Temple University, 1954, 81–84).*

ready knew it was in the red and sinking, then the decision to establish a nationwide conglomerate of electric vehicle operating companies was both devious and unwise. In retrospect, therefore, given the available evidence, we can say only that as of early 1899 the EVC might have been marginally profitable and that, to all external appearances, the company was a successful and expanding enterprise.

SECOND EXPANSION: COLUMBIA AND ELECTRIC VEHICLE COMPANY, 1899–1901

In February 1899 *Horseless Age* reported that two hundred more electric cabs had been ordered; internal communications between the EVC and Westinghouse show that negotiations were underway to supply two hundred new motors for the next batch of cabs.[59] Shortly thereafter, however, rumors of the formation of an automobile trust began to circulate through the motor vehicle industry. A February 22 article in the *New York Times* claimed that a "giant trust" had been formed by Peter A. B. Widener, William Elkins, and other owners of the Metropolitan Street Railway Company, the holding company responsible for the operation of most of New York's streetcar lines.[60] Three days later the EVC was reported to have increased its capitalization from $10 million to $25 million but denied any ties to the "auto-truck affair"—a start-up company that had been rumored to be at the heart of one of the putative trust-building schemes.[61] Whether real or imagined, the scent of profits at the EVC had begun to attract the attention of the larger financial community.

A short overview of the Metropolitan Street Railway (MSR) helps explain subsequent events. The MSR and its loose cabal of directors and senior stockholders (the so-called Whitney-Philadelphia syndicate) dominated New York's private utilities and transit service providers from the 1890s until the creation of the New York State Public Service Commission in 1907.[62] William Collins Whitney, a signal figure of the New York establishment whose family name still graces New York City landmarks such as the Whitney Museum of American Art and the Payne Whitney Clinic at New York Hospital, had assembled the MSR through judicious use of his financial and political connections. As secretary of the navy in the first Cleveland administration and one of the true kingmakers of turn-of-the-century New York politics, Whitney had profited greatly by obtaining and exploiting exclusive horse-car franchises throughout Manhattan. During the 1890s, he and his partners consolidated their holdings and gradually mechanized their lines, first with steam-driven underground cables and then with electricity and compressed air.[63] Yet as late as 1897, the company had not chosen a single mode of power for all its streetcars; instead, the MSR was experimenting with a range of options for mechanization, installing electrical conduits that were also capable of accommodating cable and even running compressed air vehicles on certain crosstown routes. By early 1898 approximately one-third of the track mileage had been mechanized, with the promise of more to follow.[64] The success of electrification also encouraged the syndicate to branch into the electric utility field, and the financiers soon had interests in several of New York's leading electricity providers, including New York Edison.[65] The economics of scale and scope inhering in these sprawling entities increased profits and further enriched the members of the syndicate. All told, the last decade of the nineteenth century was very good to the Whitney-Philadelphia interests, and the spread of electrification was an integral component of their success.[66] Concerned with financial manipulation rather than operational engineering, with patents and precedent rather than specific technologies and innovation, Whitney and his colleagues saw the growth of the electric cab fleet as an opportunity to expand their public service empire to include motorized transportation over public roads.[67]

Following the same strategy that had brought them success in the electric traction field, the Whitney interests gradually acquired shares of the Electric Vehicle Company and its parent, the Electric Storage Battery Company, on the open market. Then in April 1899 Whitney and two colleagues traveled to Hartford for a meeting with Colonel Pope and other principals of the Pope Manufacturing Company. A massive scheme was hatched whereby a new company, the Columbia and Electric Vehicle Company, was formed from the Motor Vehicle Department of the Pope Manufacturing Company, the existing vehicle assets of the EVC, and a $1 million cash investment by Whitney and his partners.[68] The new company was to supply electric vehicles to a group of regional operating companies, one of which was established in each of the nation's major metropolitan areas. These operating companies, in turn, were to operate electric cab fleets and sell vehicles to the public. In exchange for this licensed right, the parent company received 20 percent of the capital stock of each operating company and 2.5 percent of gross receipts.[69] On its face the plan looked reasonable. The electric cab business in New York appeared profitable, and expanding operations and consolidating holdings were logical strategies to allow the Whitney-Philadelphia syndicate to continue to dominate transportation service in New York City. For Colonel Pope and his manufacturing operation, the compact with Whitney allowed Pope to capitalize on his limited foray into the motor vehicle field, while freeing him to devote his own energies to building the American Bicycle Company, a short-lived bicycle trust.[70] And for the other EVC shareholders (many of whom, not by coincidence, were also owners of the Electric Storage Battery Company), the deal held the prospect of thousands more electric cabs in service, each requiring a lead-acid battery pack. Moreover, each of the partners in the Columbia and Electric Vehicle Company expected that their collective hold on the motor vehicle industry would strengthen over time. By acting decisively to corner the market in 1899, all of the concerned parties hoped to guarantee markets and profits for many years. The local New York papers hailed the combination and reported that in exchange for "unlimited means," Whitney and his partners "have asked that the gradual perfection of electric carriages which might have been expected in the next five years be completed in three months. The engineers have accepted the task and have expressed confidence of fulfilling it."[71] Indeed, had the combination succeeded in establishing an electric automobile trust, William C. Whitney and Albert Pope might now be remembered as (in)famous monopolists and system builders like John D. Rockefeller and Thomas Edison.

Even as Whitney boarded the train to return to New York, a number of fateful changes had been set into motion. By early May 1899 the promised Columbia and Electric Vehicle Company was incorporated in New Jersey with the stated aim of producing electric vehicles for exclusive sale through the EVC to its regional operating companies. The central station of the EVC operation in New York City was reincorporated as the New York Electric Vehicle and Transportation Company, and similar entities were later created in Boston, Chicago, Rochester, Philadelphia, Mexico City, New Jersey, and Washington, D.C. Provisional arrangements were made for companies to be created in every state of the union.[72] By the end of the month, the various companies within the EVC fold had been authorized to issue stock valued at more than $100 million, an astronomical sum in 1899 dollars.[73] In Boston and Chicago, leading figures in the electrical industry such as Charles Edgar and Samuel Insull were recruited to the cause.[74] Also in May, in connection with the annual convention of the National Electric Light Association (NELA), a fifty-one-car parade of electric vehicles, including twenty-five EVC vehicles requisitioned for the day by Fred Vieweg, traveled from Madison Square to Grant's Tomb and Columbia University, where guests were served tea and received by G. F. Sever of the electrical engineering faculty. Despite outright hostility from the Central Park Board of Commissioners and various logistical hurdles erected by local police, the outing was deemed a success: "the triumphant march of that long line of elegant carriages out and home, under conditions about as severe as they could be made, was an event full of significance and gratifying to all who participated in it."[75] For many of the participants, including wives of NELA members, it was their first ride in an automobile, and its success seemed to bode well for the fate of the EVC venture.

To say that the EVC was the most legitimate early effort to consolidate the nascent motor vehicle industry is damning with faint praise. Many other plans never made it past the gossip columns. A few firms, such as the Anglo-American Rapid Vehicle Company and the General Carriage Company, were actually incorporated but never managed to put even a small fraction of their capitalized value to work in building and running vehicles.[76] Hiram Percy Maxim, in his memoir written thirty-five years after the events, described the EVC scheme as "a very broad one, promising all manner of possibilities in the way of stock manipulation" and did not venture to guess "whether it was intended to develop profits out of earned dividends or by unloading the stock on the public."[77] The lead editorial in the May 3, 1899, issue of *Horseless Age* inveighed against the establishment of a "Motor Vehicle Trust," arguing that "the period of the trust has not come yet. . . . Any trust that is undertaken at this early day will be a good thing for investors to let alone."[78] But with rumors of one enormous scheme after another running rampant, caution was not the order of the day.

Back in Hartford, meanwhile, the former Pope Motor Vehicle Department faced a considerable logistical and engineering challenge. Whitney's financial infusion was a down payment on 1,600 electric cabs, and the industrial base of the Connecticut River Valley swung into action to fill the order. Hermann Cuntz, a patent attorney employed at the Pope Manufacturing Company, described the influx of engineers into Hartford as "the talk of the town" and reported that downtown hotel rooms were fully booked for months. William Hooker Atwood's New Haven Carriage Company was purchased outright, and other supplier networks were created to satisfy demand.[79] In June the EVC contracted with Studebaker Brothers of South Bend, Indiana—the largest carriage-making firm in the country—for the production of one hundred vehicle bodies. In July the company discussed expanding the original order from 1,600 to 4,200 vehicles.[80] Clearly, the new conglomerate mobilized considerable resources from the spring of 1899 on; but just as in 1897, when demand for vehicles had increased by one order of magnitude with the initial creation of the EVC, the company failed to take advantage of the opportunity to transform the production process. Published reports highlighted the continued use of "carriage makers' methods throughout" in keeping with the view that EVC customers "are a conservative class and scrupulously avoid anything that suggests a departure from the appearances to which they are accustomed." And while the company touted its development process and claimed that its designs were approved only after surviving an internal "furnace of criticism," in practice standards proliferated, and no coherent policy emerged about how to procure component systems and parts. Vehicles were produced with single and dual motor configurations; solid and pneumatic tires were employed. Some iron work was done in Hartford by hand, but springs and motors were bought from outside suppliers.[81] Meanwhile, in Chicago the company simply bought an existing factory owned by Seimens and Halske and converted it to produce components for electric cabs.[82] Rather than striving to become an integrated manufacturer, the EVC remained a mass purchaser of vehicle components manufactured by many different suppliers. And as part of its effort to control competition, the company overpaid to acquire existing facilities from competitors rather than integrating and consolidating effort into a single line of production. Instead of building the company by gradually expanding production in line with sales, the conglomerate paid out more and more capital to devour its competitors and then compounded its problems by not bothering to digest them.[83]

Transfer prices among the various holding companies also underscore the extent to which the conglomerate failed to act as a unified entity. Account books from several of the regional operating companies suggest that the parent companies (the Electric Storage Battery Company, the Electric Vehicle Company, and Columbia and Electric Vehicle Company) might have kept prices deliberately high to siphon money up the corporate food chain. Because the regional operating companies were captive consumers of the EVC products and were required by their charters to pass along a fixed percentage of revenues to the central holding companies, the original members of the investment syndicate made money both when the vehicle was transferred to the operating company and when that vehicle was sold to the public or placed in service in a local cab fleet.[84] To be sure, there was always an incentive to lower production costs, but making money through a stock pyramid scheme was easier and more familiar for the financiers than actually figuring out how to produce more with less.

HORSELESS AGE DRAWS BLOOD, 1899–1900

The establishment of the Columbia and Electric Vehicle Company and its subsequent effort to corner the electric vehicle market coincided with a series of inauspicious reports on the plights of other electric vehicle ventures. In mid-August 1899 *Horseless Age* announced that the London Electrical Cab Company had shut down, claiming that it could not retain its employees: once trained, the drivers were soon hired away by private vehicle owners needing chauffeurs and offering better working conditions.[85] Later that month, the magazine reported that an electric cab experiment in Paris was being restructured. The French electric cab company had stopped offering public cab service. From then on, vehicles were available only for private, prearranged trips.[86] The public image of the electric vehicle suffered another blow when the Automobile Club of France announced initial results from its summer endurance test of electric storage batteries: of the twenty-three different batteries tested, four had already failed, and four more were on the verge of being withdrawn.[87] That week, an editorial in *Horseless Age* concluded that recent events "prove what might have been known without the waste of so much money—that in the present state of the storage battery art, electricity cannot compete with horses in public cab service, but is of necessity confined to luxurious use."[88] Regardless of whether the European operators had truthfully reported the reasons for stopping their electric cab services, the announcements produced a change of heart at *Horseless Age*. The combination of a booming American trust and broken European cabs led the magazine to take a stand.

One month later, the press assault began in earnest. As *Horseless Age* began its third year of publication, it issued a "declaration of independence," promising to take a more aggressive stance against "promotions and fakes which stand in the way of progress."[89] Although the magazine later weighed in on other controversial issues as well, the EVC cab scheme bore the initial brunt of the journal's editorial zeal. A front-page editorial titled "Electric Vehicles and Their Limitations" noted weight, cost, and delicacy as potential barriers to the widespread use of electric vehicles for general-purpose transport service and observed that electrics had been "banished from the broad field of the workaday world to the more limited field of luxury and incidental uses." But the journal saved its strongest words for the storage battery at the heart of the electric vehicle system: "No modern invention has enlisted so large an expenditure of time and money with so little result as the electric storage battery. Fortunes have been wasted in fruitless efforts to overcome by some mechanical means the inherent weaknesses of the storage cell."[90] The article failed to mention the "inherent weaknesses" of the competing technologies, nor did it acknowledge existing, profitable applications of electric storage batteries; but this attack was tame compared to what followed.

The lead-acid battery became the lightning rod for public criticism of the electric cab enterprise. The EVC conglomerate was derisively called the "Lead Cab Trust," a moniker that stuck.[91] When engineer Elihu Thomson, in a public lecture, announced his preference for steam power and described the electric storage battery as an "unmitigated nuisance," *Horseless Age* editorialists adopted the phrase and sprinkled it throughout their articles.[92] And when proponents of the EVC tried to defend their company and their technology, *Horseless Age* would have none of it. In October 1899, EVC chief engineer Maxim wrote an article in *Cassier's* praising the accomplishments of the electric vehicle on both sides of the Atlantic and noting the difficulty of operating a gasoline vehicle. *Horseless Age* promptly accused Maxim of being "a blind and infatuated partisan of the lead wagon" who was hiding that "sick baby, the storage battery."[93] An article titled "Storage Battery Financiering" concluded:

> [The financiers'] millions cannot change one atom of the storage battery or make it anything but what it is—a heavy, inefficient, delicate and destructible apparatus wholly unfit for general locomotion. . . . Nor will they be able to change the nature of the rubber tire and render it more suitable for sustaining heavy weights and resisting the wear and tear of rough roads. No . . . , the limit has been reached. Rubber is rubber and lead is lead. . . . The outlook for the lead cab is utterly hopeless.[94]

A brief and mysterious article titled "If" followed: "If cows were physically and mentally different they might manage two tails conveniently. . . . If we would make improvements we must improve things as they exist. The new is conditioned upon the old."[95] Again, the attack on the inherent limitations of the electric storage battery continued, although in the case of the "If" article, we must read between the lines to see its intended target.

After several months, the attacks on the storage battery softened, in part because the EVC scheme appeared to be unraveling, in part because *Horseless Age* saw that it had overstepped the boundaries of good journalism. A January 1900 editorial summarized the magazine's "reform and education" achievements and claimed to have substituted "the sober judgment of common sense" for "the rosy visions of the promoter."[96] The journal was prepared to separate the general issue of electric vehicles from that of the Lead Cab Trust and defended its previous attacks as being necessary to save the electric vehicle from its unscrupulous promoters:

> Had this scheme [the electric automobile monopoly] been carried out the industry would have been crippled for several years to come. . . . the electric vehicle itself would have received a setback from which it could not have recovered in years. . . . When the Lead Cab Trust is entirely through promoting and the electric vehicle stands on an honest basis, the *Horseless Age* will take it up again from an engineering point of view and endeavor to repair the damage promoters have done it, defining more carefully the limited field it is possible for it to fill in the new locomotion.[97]

The magazine professed, in other words, to have the best interests of the electric vehicle at heart: they killed it to save it. But other observers were not convinced. Another leading auto journal, *Motor Age,* questioned the motives of *Horseless Age* and accused editor-in-chief E. B. Ingersoll of being an "autoelectrophobe." Disclaiming any personal or financial interests in the matter, the *Motor Age* editorial spoke of its "duty to its readers . . . who may have been misled through the untruthful and vicious diatribes, which, in his capacity as editor, he [Ingersoll] has seen fit to foist on an unsuspecting and . . . ignorant public."[98] Editorialists at *Electrical World,* although willing to confess their "predilection . . . writ large on our title page," were similarly perplexed, characterizing Ingersoll's outburst as "an extraordinary jumble of inaccuracy and misinformation."[99] By early spring 1900 the editorial policy of *Horseless Age* had mellowed, admitting of all three motive powers that "each has its peculiar advantages and sphere of use, and the first thing to do is to recognize frankly its limitations, and to set about perfecting it within those bounds."[100]

In the end it is hard to know what exactly prompted Ingersoll to take such a dim view of the electric vehicle when he did. To be sure, the electric storage battery of 1899 had its problems, but it could not have been as bad as Ingersoll wanted the public to believe; otherwise, no one would have bothered to pursue its development. As it was, several independent electric vehicle transportation projects were proposed during and immediately after the magazine's flurry of criticism.[101]

THE EVC REGIONAL OPERATING COMPANIES

The experiences of the local operating companies suggest that *Horseless Age* was both right and wrong in its assessment of the Lead Cab Trust. On the one hand, the company endured a stretch of operational problems and bad luck. Vehicle production and delivery were well behind schedule. Promises had been made and capital raised to finance the production of as many as 10,000 electric cabs; yet as the year and century drew to a close, no more than several hundred vehicles were in service nationwide. Despite the slow pace of vehicle production, a 2 percent stock dividend had been declared in late 1899, and financial manipulations continued. During Ingersoll's protracted press attack, share prices of companies connected with the EVC scheme were falling, perhaps as news of various operational problems leaked into the market. During December 1899, however, speculation in the stock of an unrelated monopoly scheme called the General Carriage Company drove its share price from about fifty dollars to two hundred dollars overnight. The General Carriage Company claimed a special franchise allowing it to operate electric vehicles for hire on any street in the state of New York, but other sources suggested that the franchise was unenforceable and therefore valueless. The "air power interests" (individuals, including members of the Whitney syndicate, aligned with schemes to use compressed air as a power source) were also reported to have invested in the General Carriage Company. Financial insiders suggested that the soaring stock price was not due to changes in the condition of the business but to other, purely speculative factors: "the advance was ascribed to the existence of a short interest, with no floating supply of stock for the shorts to borrow to make deliveries with." In other words, the share price had soared as a result of a frantic bid to cover short positions. By late December the speculative bubble had broken, and by early January the bid price for General Carriage had fallen to one dollar per share.[102]

Many other stocks were drawn along in the collapse of the General Carriage Company. *Horseless Age* reported that electrical stocks "fell with a thud."[103] The financial crisis could not have come at a worse moment for the EVC, for a factory fire in New York on December 19, 1899, had destroyed eighteen new vehicles ready for delivery and nine more being repaired for the New York branch.[104] Although the company had already distributed forty of the new hansoms to local transportation companies in Philadelphia, Chicago, and Boston, initial reports from the field were not favorable. The cabs had been rushed into service, and "nothing on the vehicles was right."[105] To be clear, there was no evidence of any direct financial link between the General Carriage Company and the EVC, nor were the difficulties encountered by the EVC necessarily harbingers of imminent financial disaster. Nevertheless, troubles at the General Carriage Company carried over into the rest of the electric vehicle sector.[106] Underscoring this point, during 1900 some of the shares of the EVC operating companies—in particular, those of the New England Electric Vehicle Transportation Company—were undervalued by the stock market in apparent reaction to the trying events of December 1899. At several points, EVC representatives noted that company coffers contained more cash than the company was worth at prevailing share prices. In the end, the General Carriage Company stock crisis, although disastrous for the short-side speculators who were forced to cover their positions, pressured EVC stocks lower but did not precipitate a complete collapse in EVC shares (figure 2.7). As was often the case, share prices did not necessarily reflect operational reality. While the parent company rode out the crises of 1899 and 1900, the regional operating companies began to breathe life into the idea of electric passenger service outside of New York City.

Boston and Newport

The New England Electric Vehicle Transportation Company launched cab service in Boston in May 1899 but focused its initial efforts in Newport, Rhode Island. There, for the first time, the company offered to rent electric vehicles with or without drivers. For the modest sum of $150, the customer was entitled to drive six hundred miles per month.[107] An assistant engineer described the situation: "at the start there were three widely differing factors that had to be brought into harmony; namely, an empty carriage and horse barn, the summer colony and the automobiles." None of the three elements was ideally adapted for the other two. The facility was small, poorly laid out, and under construction throughout the summer. Among the residents, the twenty to thirty vehicles were enthusiastically received; at times, the waiting list for vehicles was thirty names long, an indication that perhaps the company was undercharging for its services. Most customers preferred to operate the vehicles themselves, requiring a team of paid drivers to fan out across the town at 7 A.M. delivering

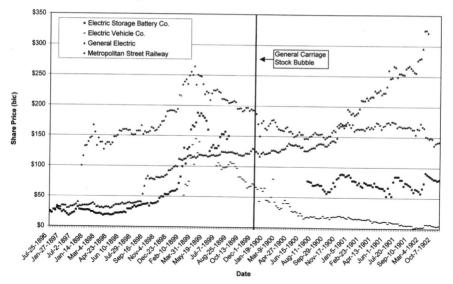

Figure 2.7: ESB,EVC,General Electric and MSR

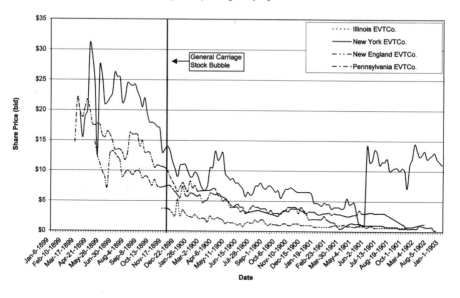

Regional Operating Company Shares

FIGURE 2.7. Weekly share prices of select electrical stocks, 1896–1903. Mark indicates stock crisis of the General Carriage Company in December 1899. Note that prices of EVC-related companies—especially the New York Electric Vehicle Transportation Company—maintained some value after the correction following the collapse of General Carriage. *Source: Collected from "Financial Intelligence" column in* Electrical World *and from* Commercial and Financial Chronicle, *1896–1903.*

cars for the day. For their part, the renters showed little care in handling the electric carriages. Reports of reckless driving were common, and there were many stories of vehicles running into barns, through bonfires, or down staircases. An emergency wagon was often dispatched only to discover that the customer had failed to insert the safety plug, the practical equivalent of forgetting to turn the power on.[108] Notwithstanding these problems, an electric vehicle parade was organized in early September with the help of the local station. Several dozen vehicles were decorated with flowers and lights, providing a spectacle for the entire town.

At the close of the summer season, the company began to expand operations in Boston. By November only eight of the nineteen new vehicles shipped there were in working condition, prompting EVC general manager Johnson to dispatch consulting engineer William Kennedy to the scene for a firsthand report, which is instructive on several counts.

Kennedy spent ten days in Boston assessing the situation and helping return vehicles to service before filing his report on November 21, 1899. He attributed "most of the trouble at [the Boston] station" to two related causes: "first . . . a total absence of a regular daily inspection of the mechanism of the cab, no precaution being taken to guard against the effects of ordinary wear and tear . . . and second . . . a lack of proper lubrication of the motor and wheel bearings and the moving parts of the brake and steering gears." A vehicle was inspected only if it ceased to function entirely, and almost half the motor and wheel bearings (56 out of 120) contained no lubricating oil at all. Often, where oil had been applied, it had been done so improperly, resulting in several breakdowns due to overheated bearings. Kennedy also noted that vehicles were arriving from the factory improperly assembled. The wheels on five of the new cabs were not properly aligned, resulting in reduced range and performance; and the vehicle controllers exhibited "inherent defects." Chief engineer Condict would later note similar problems with the vehicle controllers and proposed minor modifications to limit sparking and improve mechanical performance.[109] Although Kennedy was eventually able to return nine of the eleven inoperative vehicles to service, the episode did not bode well for the future of the company's plan to provide mass-produced electric vehicle service.

At first glance, therefore, the attacks in *Horseless Age* seem justified by the cabs' poor performance on the streets. But whereas

Horseless Age attacked the physical properties of lead and rubber, the Kennedy report highlighted the importance of high-quality production, maintenance, and training to the success of the enterprise. Kennedy asserted that the drivers were "too young and quite frequently reckless" and that the controllers suffered from "rough, abusive usage in the hands of the drivers." Nevertheless, he absolved the drivers somewhat by noting that the "severe treatment" was partly the result of "neglect of the attention necessary to keep them [the controllers] in proper working condition." The report concluded that the company should have dispatched a mechanic "familiar with the various parts and the proper care of the vehicle" to train the local drivers and station mechanics. Subsequently, EVC management prepared a vehicle manual with "full instructions for care and maintenance" and agreed to begin sending "an experienced man . . . to the vehicle stations to instruct in handling and inspection."[110] In other words, among EVC midlevel managers, engineers, and mechanics, the problems had nothing to do with lead and rubber but with the integration of human and technical components into a working transport system.

These were problems that could, in principle, be solved. By the summer of 1900, the New England Electric Vehicle Company was operating approximately 175 electric vehicles, 100 in Boston and the rest in Newport. Net profits of $10,000 were reported from the Newport station, indicating gross margins in the neighborhood of 22 percent.[111] Meanwhile, the design of the Boston Cyclorama station, overseen by general manager and chief engineer Knight Neftel, included several important innovations. In addition to the hydraulic positioning and battery exchange system developed by Condict, the Boston station took advantage of a mechanical hydrometer that increased the speed at which the specific gravity of the electrolyte could be tested and an internal power trolley that allowed vehicles to move about inside the station without batteries. Aiming to expand the effective operating area of the company's electric vehicles, Neftel planned to establish concentric rings of charging stations at increasing distances from the city hub. As of June 1900, half of the first ring was complete. Remote charging stations were available in Cambridge and Chestnut Hill, with Dorchester and Chelsea expected to follow; and plans for the second ring had already begun.[112] In addition to cab service, the station supported a range of other electric vehicle activities such as commercial truck rental, local electric bus service, and private car rental and maintenance.[113] Although the stock price of the New England company languished, by August 1900 Neftel was able to report that it was meeting its payroll and expenses and that the fleet was covering more than 2,000 miles per day. Two months later, an anonymous company official confidently predicted that the company would reach profitability of 5 to 6 percent per year for fiscal year 1900–1901.[114]

New Jersey

In Atlantic City, where the New Jersey Electric Vehicle Transportation Company operated a small fleet of electric buses and cabs, a similar story unfolded. In late 1899, EVC general manager Walter H. Johnson dispatched sales agent J. M. Hill to investigate the prospects for expanding electric vehicle service in New Jersey, both inland to Philadelphia and down the shore to Cape May. Several private customers had expressed their desire for convenient, reliable charging facilities en route from New York to Philadelphia, and the growing tourist traffic among the shore towns also looked promising. In mid-November 1899 Maxim and a colleague, in a well-publicized demonstration of the capabilities of electric vehicles, traveled from Atlantic City to Philadelphia on a single charge. Later that month, Hill reported that, although post road service from New York to Philadelphia was desirable, the roads were often impassable for weeks at a time during the winter, leading him to recommend summer stations both on the post road (at Plainfield, Princeton, Trenton, New Brunswick, and Elizabeth) and along the shore (at Seabright, Lakewood, Long Branch, Allenhurst, and Spring Lake). Hill doubted that the factory production schedule could be counted on to deliver any eight-passenger omnibuses in time for the 1900 summer season, but he was nonetheless optimistic about the general prospects for the electric vehicle business in New Jersey.[115]

Early in 1900 Johnson sent Frank C. Armstrong, a new sales agent, and EVC engineer C. Gilbert to arrange station operations and begin negotiating with potential customers. The two men recommended Atlantic City as the main base for the Jersey shore service: "Atlantic City, of all the cities on the coast, is most likely to show good results."[116] The town was already popular with convention-goers and boomed with middle-class tourists during the summer season.[117] To begin, the EVC planned to offer electric omnibus service from the train station to local hotels. The electric buses would not do away with the need to keep horses for hauling and deliveries; but from an advertising and marketing perspective, those hotels that signed on would undoubtedly distinguish themselves from their old-fashioned competitors. And because Atlantic City thrived on novelty—even oddity—the electric bus service would attract considerable interest. By March, Armstrong asked Johnson to come to Atlantic City in person to "positively close [deals] with six hotels at once," thereby securing the sale of ten buses.[118] Armstrong hoped to be ready to inaugurate omnibus service by the time the tourists started to arrive in May, but protracted negotiations and production delays postponed the official opening until June.

During its single year of operation, the New Jersey Electric Vehicle Transportation Company provided a range of different transportation services. Private owners of electric vehicles could have them garaged and maintained for seventy-five dollars for the three-month summer season. An extra five dollars per month covered unlimited recharging at any one of the company's service stations along the shore. In Atlantic City proper, the company also offered electric hansom cab service, omnibus service for a select number of hotels, and limited public transport service up and down Virginia Avenue. In some instances, the company would rent its buses to private customers for special functions. The New Jersey operation is also the only piece of the EVC corporate puzzle whose complete operating records have survived. We know how many miles each omnibus and cab traveled, can be reasonably sure of the revenue they produced, and have a good idea about how much it cost the company to generate this revenue.

The service records show that the New Jersey EVC branch got off to a slow start. Although Hill, Armstrong, Gilbert, and others worked diligently throughout the spring of 1900, by the time the company finally opened its doors in June the central station was not yet fully operational, nor had the company received all the vehicles it had been promised. Total mileage in June (1,345 miles) was less than one-quarter of the monthly totals recorded later in the season (5,198 miles and 6,309 miles in July and August respectively); and a number of the omnibuses showed very low individual mileage, indicating that they did not enter service until late in the month. Similarly, the two hansom cabs assigned to Atlantic City logged more than 75 percent of their total mileage for the month in the last week of June.[119] The station may have officially opened at the beginning of

June, but it was not operating at planned capacity until a full third of the tourist season had already passed.

Once the company was in a position to offer its full range of services, business picked up considerably. During the eleven weeks between July 1 and September 15, the company served five hotels each week with an average of six buses running each day. Total weekly mileage averaged 740 miles (515.5 miles paid). Each omnibus covered 17.5 miles per day (total), about 70 percent of which (12.2 miles) generated revenue. Hansom cab mileage also rose during this period, averaging 18.49 miles per day, 12.48 paid (67 percent). Likewise, private bus rental and public transit service along Virginia Avenue peaked. The weekly reports do not indicate how many vehicles were assigned to this route, so it is not possible to calculate daily mileage. These two and a half months marked the brief heyday of the venture.[120]

By mid-September, however, service had declined precipitously. Not only had the tourists gone home, but nearly the entire vehicle fleet—hansom cabs and omnibuses—was out of service. In contrast to the situation in Boston, where the vehicles malfunctioned almost from the day they arrived, in New Jersey the hansom cabs and omnibuses were delivered, rushed into service, and appeared to be yielding reasonable if unspectacular results. Not until September did the liabilities of the electric transportation system catch up with the modest facility. Of fourteen new sets of batteries originally delivered to the operating company, ten could no longer be used, and Walter Johnson dispatched several underlings from headquarters to investigate the situation. As in Boston, investigators found that the vehicles had not been receiving adequate and regular maintenance. The instruction manual prepared in the aftermath of the Boston fiasco had apparently been ignored: "No records of any kind have ever been kept, and no attention has ever been paid to the instructions that have been furnished in regard to the treatment of batteries, their inspection or care."[121] Batteries were watered with acid instead of pure water. Vehicles were driven too hard. Improper lubrication resulted in frequent hot bearings.[122] So went the litany of woes. One inspector posited that because all the batteries failed in the same manner (the negative plates had shed their active material), the fault might have been with the battery plates rather than the maintenance procedures of the Atlantic City staff. H. Crittenden wrote that he "was inclined to lay most of the fault on the negative battery plates themselves—

they are a summer output which have always proved to be inferior to winter ones."[123]

Two types of problems, however, were unique to the Atlantic City station and suggest the broader set of barriers facing the electric vehicle in semi-urban America. First, the EVC observers reported that power from the electric utility station could not be counted on. Line voltage was irregular, and power was often unavailable during off-peak hours. Buses would return to the station house with completely empty batteries only to "stand discharged until 12 midnight and sometimes later before being charged, as there was no current furnished by the Electric Light Company between 6 p.m. and 12 midnight." Leaving the batteries discharged not only shortened their service life but also prevented them from receiving a full overnight charge. By 6 A.M. the vehicles were needed to meet the early trains. Schedule permitting, the batteries were then given short overcharges during the day. On other occasions, when power would fluctuate overnight or cut out altogether, the vehicle batteries discharged back into the utility line. Drivers arrived for work in the morning to find that the vehicle batteries had been carrying the overnight lighting load for the town and were completely exhausted. In short, the New Jersey station typified the problematic relationship between the central station and the prospective electric vehicle charging garage. It is evident that although vehicle charging was an attractive business for a local lighting company (it occurred mostly off peak and therefore did not require expanding capital stock), the central station in Atlantic City did little to help the EVC service station (see chapter 3). The supply of electric current for the vehicle station fluctuated and was unreliable. The company's only alternative would have been to install an isolated generating plant, but this option would have been prohibitively expensive and would likely have led to a separate set of operational problems. Without active support from the local utility, the electrical infrastructure could not provide adequate service to the fledgling enterprise.

Charging problems were compounded by a second problem at the Atlantic City station: the rheostats used to charge the batteries were difficult to keep in service. With power available only for limited periods of time, all the batteries had to be charged at once. If one rheostat was not working at night, one less vehicle was available for service the next day. Because the rheostats could not be repaired locally, a burnt-out rheostat was useless until it was returned from the manufacturer. Unlike other design and maintenance problems that were at least technically of the company's own making, the difficulties associated with vehicle charging were beyond the effective control of the Electric Vehicle Company and its Atlantic City station manager.[124]

Ultimately, however, it did not matter who was at fault. The vehicles were poorly maintained and inadequately charged. Whether the batteries went bad or were made bad was of little consequence to the EVC managers in New York. Reviewing the situation with his colleague, Johnson resigned himself to the inevitable: "It looks as though all the money we made and a few hundred per cent more will be spent in repairs. You and I must get together on this matter, also regarding the closing of the [sub] stations."[125] The company spent the next several months trying to stem its losses, repair its vehicles, and slowly rebuild its service from a weekly low of ninety-six miles of total service (sixty-eight miles for one omnibus and twenty-eight miles for one half-time hansom cab) in early November 1900. Frank C. Armstrong, by then one of Walter Johnson's trusted lieutenants, was sent to Atlantic City on October 15 to personally untangle the affairs of the troubled company. Armstrong promised to "make payroll during the slack season," and records show that the Atlantic City station posted an operating surplus after December 1, 1900 (figure 2.8).[126]

By early 1901 New Jersey service was on the rise, and in April—in the final weeks before the station was closed—the Atlantic City fleet was covering nearly as many miles as it had during its peak the previous summer. In New England, too, the situation appeared to be turning around. Including buses, trucks, cabs, and long-term leases, 230 electric vehicles were in service, several charging substations had been established, and the electric cabs were responding to approximately two hundred calls and covering more than seven hundred miles per day. A lead editorial in *Electrical World* touted the accomplishments as "most laudable" and praised the "courageous investment" and "infinite amount of hard work and ingenuity" that had gone into building the EVC fleets. The article enumerated almost 1,000 EVC vehicles in active fleet service in cities as far apart as Berlin and San Francisco and as dissimilar as Atlantic City and Mexico City. "Electric automobiles," the editorial concluded, were "quietly making headway."[127]

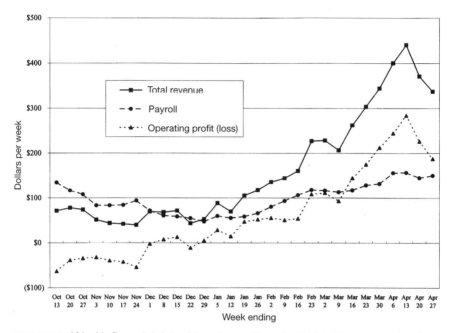

FIGURE 2.8. Weekly financial data, New Jersey Electric Vehicle Transportation Company, Atlantic City Station, October 1900–April 1901. *Source: Frank C. Armstrong Papers, Henry Ford Museum and Greenfield Village, Dearborn, Mich., box 6, folder 2, "New Jersey Electric Vehicle Transportation Company."*

The Collapse of the National EVC

Starting in late February, however, EVC management opted to close many of the local operating companies outside New York. The Illinois Electric Vehicle Transportation Company was the first domino to fall. When the Chicago cab drivers staged a strike rather than accept a profit-sharing plan proposed by management, President Samuel Insull chose to liquidate the company, claiming that "local conditions" and "the consequent high cost of maintenance" had rendered the service unprofitable.[128] An editorial in *Electrical World* suggested that the decision was "a discouragement rather than a surprise to those who have been watching the situation"; poor road conditions, regulated fares, competition from Chicago's "most comprehensive" streetcar and elevated system, and the disproportionately industrial populace had limited the prospects for the venture from the start.[129]

Several weeks later, the New England branch followed suit. Although the company was well capitalized, the directors felt it was unwise to expend more resources in operations and "in the purchase of other vehicles" where "a return cannot be obtained commensu-

rate with the amount of capital."[130] Again, *Electrical World* defended
the general idea of urban electric vehicle service: "we have never been
accused of undue optimism regarding the storage battery and have
tried to preserve a thoroughly conservative view of its possibilities,
but, frankly, we do not believe that there is any inherent reason why
the electric automobile should not be entirely successful, particularly
in its chosen field of urban service. . . . It has already scored some really wonderful performances, and we should be very sorry to see the
reduction of unprofitable service here and there charged against it as
a failure." The editorialists instead attributed the decision to abandon the electric cabs to the high quality of American streetcars: "because street car service is so cheap, so good, and so ubiquitous . . . the
public [has come] to look upon cab service not as an every-day necessity but as a luxury of dubious utility." They also cited the poor
behavior of the cab drivers: "if the [streetcar] conductors were abusive and drunken and held up their passengers for half a dozen times
the regular fare in case of rain or if there were ladies in the party at
such times, street cars would soon come to be in evil repute."[131]
Finally, in May the EVC also closed the Atlantic City station and recalled Armstrong to the New York office. By 1901 the major share-
holders of the vehicle operating companies were more concerned
about preserving capital than with the ultimate success or failure of
the electric cabs themselves. Independent of financial considerations,
electric vehicles required vigilant and labor-intensive inspection and
maintenance to provide any service at all. Neftel and Armstrong had
done their parts, but the EVC senior management had decided it was
time to retrench. Nevertheless, at the time that service was terminated
in Atlantic City and, probably, in Boston, the electric fleets were operating in the black and might have been able to survive. Only two
short years after the historic meeting in Hartford between Whitney
and Pope, the Electric Vehicle Company had effectively abandoned
its expansive vision of Columbia electric cabs in every American
city.[132]

New York

Despite the problems at the EVC operating companies, the fate of
the original New York station suggests that, as of late 1901, the general prospects for electric vehicle transportation were not nearly as
dismal as has been traditionally assumed. In late fall 1901 Walter
Johnson described a massive shakeup at the EVC. The old Riker

factory in Elizabethport, New Jersey, was closed, and the corporate offices were moved to Hartford. The Pennsylvania Electric Vehicle Company (the Philadelphia operating arm) was acquired by the ESB; and after a fivefold capital reduction, the flagship New York Electric Vehicle Transportation Company was reincorporated in December 1901 as the New York Transportation Company.[133] Clearly, the business had not developed according to the plan first proposed by Whitney and Pope in 1899. Even so, the corporate reorganization and the decision to abandon the remote operations was not a blanket capitulation. At least in New York, the New York Transportation Company continued to operate electric cabs, buses, and sightseeing coaches in and around the city until 1912.

In fact, during the years immediately following the failure of the national cab scheme, the electric cab service of the New York Transportation Company continued to innovate and expand. In late 1901 the company began experimenting with the new Exide battery designed especially for mobile applications. Any number of inventors claimed to have produced improved, lighter, more efficient, more durable batteries, yet by 1901 the New York cab company continued to use traditional Chloride-Manchester batteries built by the Electric Storage Battery Company. Unlike the alkaline battery developed by Thomas Edison, the ESB's Exide battery used the traditional pairing of lead with sulfuric acid. And unlike some of the other promised improvements (Edison's nickel-iron battery, initially announced in 1901, was not widely available until several years later due to production problems), the Exide was market-ready and actually delivered better performance.[134] As Transportation Company treasurer William Palmer reported in 1902, battery performance was not a problem to be addressed willy-nilly by, for instance, increasing the mass of batteries carried or decreasing the thickness of existing battery plates: "ill considered steps in any of these directions may easily result in an increased maintenance cost which will far outweigh the advantages of increased mileage." Accordingly, the Exide battery was an important but incremental advance over the traditional Chloride cell design. Whereas Chloride cells were recharged after traveling an average of only ten miles, the new Exides, placed in the same vehicles, were allowed to run twenty miles per charge. The theoretical energy capacity of the standard Exide was only 50 percent greater than the Chloride's; but given concerns about maintaining a "relatively high margin of safety," this difference translated into an effective doubling of mileage per charge. The Exide thus represented a considerable improvement over the original batteries, prompting Palmer to conclude:

> The company is demonstrating in its daily business that the electric automobile has been developed to the point where, in reliability, radius of action and operating costs, it meets all the requirements necessary to commercial success, and that the development of a storage battery especially adapted to work under the rigorous conditions imposed has contributed greatly to the result. So far from being an insuperable obstacle to the success of electrically propelled vehicles, the battery is today better adapted to its work than some other portions of the equipment, and in point of maintenance cost, a smaller factor in the total expense of operation than are, for instance, the rubber tires on the vehicle it runs.[135]

Palmer may have been indulging in a bit of hyperbole. Expense reports from 1903–8 indicate that battery costs continued to exceed the cost of tires by a considerable margin. In his defense, however, tire costs also declined during this period as the company adopted solid tires for all its vehicles.[136] In any event, the benefits of the Exide battery were real. Its impact on the profitability of the cab operations was probably significant, and it soon became the new standard.

The relatively compact and isolating geography of Manhattan was well suited to the capabilities of the electric cabs. The large station on the West Side continued to serve as the base for cab operations, although the company eventually acquired substations and standing privileges at a number of remote locations, including the Hotel Astor (44th Street), Café Martin (26th Street), and the 34th Street ferry terminal.[137] Assessing the general public view of the service, William Palmer observed that "at rates equal to or in excess of the rates charged for the highest class of horse livery service, far more applications for service are received daily than the company, even with its increased equipment, is able to accept."[138] The following year, the company reported that "business . . . showed a substantial growth and the earnings are showing a satisfactory increase, the profits during the winter months being very gratifying."[139]

In 1904 Richard W. Meade assumed the leadership of the company. He was a capable administrator who understood the challenges of operating a privately owned, public-access transportation service company. He served as general manager and president of the New

York Transportation Company for almost fifteen years and went on to a distinguished career as president of the Detroit Motorbus Company and of the People's Motorbus Company of St. Louis.[140] In 1907 Meade ordered "upward of 100 of the company's cars" fitted with imported taximeters, thereby making the company the first to successfully operate taximeter-equipped vehicles in the United States.[141] Meanwhile, the company added a charging station on the East Side, at Third Avenue and 66th Street, and acquired the former offices and station of the Vehicle Equipment Company on 27th Street and Ninth Avenue, closer to the business heart of the city.[142]

Each of these new depots served important segments of the transportation market. The downtown station catered to the expanding commercial vehicle market (see chapter 4). At the uptown station the company stored and maintained the large share of its electric vehicles that wealthy New Yorkers leased by the month or season. In exchange for a fixed monthly sum varying from $275 to $375 depending on the specific type of vehicle, patrons were provided with their own private car and driver. For contracts exceeding four months, the customer was allowed to affix a crest or a monogram to the door of the electric carriage. Payment entitled the renter to call on the driver for no more than twenty-four hours per forty-eight-hour period and included 650 miles per month. As the winter season drew to a close, the company shipped many of these vehicles to Newport, where a similar service was offered during the summer.[143] As late as 1906, this line of business was both popular and profitable. The company also stored and maintained privately owned vehicles and in 1905 used the East Side facility to introduce a small number of gasoline touring vehicles for rental.[144]

The New York Transportation Company also operated several subsidiaries, including the Park Carriage Company, the Metropolitan Express Company, and the Fifth Avenue Coach Company. Each daughter company focused on a different segment of the transportation market. The Park Carriage Company operated horse-drawn and electric vehicles in Central Park and scheduled electric touring coaches from midtown hotels to upper Manhattan landmarks such as Grant's Tomb and Riverside Park. As of mid-1906, the Park Carriage fleet consisted of thirty-one horse-drawn carriages, eight electric vehicles, and two experimental gasoline cars. Metropolitan Express provided commercial package service using special streetcars that ran on the Metropolitan Street Railway lines starting in 1901. Little is known about

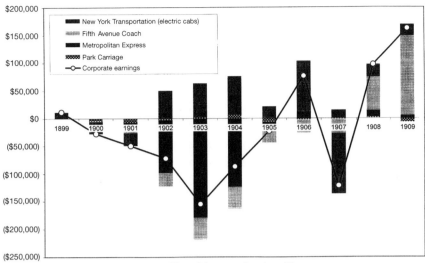

Financial year ending June 30

FIGURE 2.9. Contributions of subsidiaries to net earnings of the New York Transportation Company, New York City, 1899–1909. Cab operations produced positive earnings from 1902 to 1909 except for the October 1906 strike year (fiscal year 1907). *Source: Richard W. Meade Papers, Columbia University, New York, box 15, folder "New York Transportation Company, Statement of Income and P&L."*

the details of its operations, and in July 1904 the subsidiary was leased to an outside operator. As shown by its combined financial results, the express company was a persistent drain on the treasury of New York Transportation.[145] As figure 2.9 summarizes, before fiscal year 1907 only the electric cabs regularly produced an operating surplus.[146]

As time passed, however, the electric vehicles looked increasingly out of date. New cabs were difficult to procure, and finally several unforeseeable events effectively crippled the electric fleet. First, a protracted strike by cab drivers in October and November 1906 resulted in heavy losses. Seeking union recognition, a closed shop, and an increase in the number of full-time drivers, approximately 250 of the 325 regular drivers fought President Meade and the Transportation Company for almost four weeks. In response, the company arranged to house and feed two hundred men, hired strikebreakers, and organized a forty-person bicycle security detail to accompany selected cabs as they traveled the streets. Violent conflicts were common, and at least twenty-five people were arrested "for various forms of assault and riot."[147] In the realm of labor relations, the New York Transportation Company remained closely tied to its roots in the streetcar industry.[148]

Compounding the damage from the strike, a massive fire at the central station in January 1907 destroyed two hundred electric cabs.[149]

As the New York Transportation Company sought to rebuild, the old EVC was teetering on insolvency, the national economy was reeling from the Panic of 1907, and across the Atlantic gasoline cabs were demonstrating advances. Accordingly, despite complications and added expense, Meade chose to import a batch of fifty gasoline Delahaye landaulets from France. Along with Harry Allen's New York Taxicab Company, which also chose to import cabs from the Continent (from Darracq), the Transportation Company was among the first firms to operate gasoline taxicabs in the United States.[150] The American automotive press criticized the decision to import; but Meade, like Allen, believed that American-made vehicles were not durable enough to withstand the punishment of daily cab service. Within several years, American manufacturers had imitated European practice, and by 1909 domestic manufacturers accounted for nearly half the cabs plying the streets of New York.[151]

Beyond the problem of finding a suitable vehicle, initial efforts to introduce gasoline cab service in New York faced several other important hurdles. As had happened with the first electric cabs, demand for the new service outstripped supply; the resulting taxicab investment boom failed to anticipate the operating problems that the vehicles encountered and led to competing and disputed claims about profitability. An early account of the New York Taxicab Company fleet, for instance, reported that the cabs were yielding twenty-five dollars per day on average, with exceptional cases running to forty dollars per day. Although the drivers received no fixed salary, their 20 percent share of gross earnings was well in excess of the prevailing wage in horse or electric work.[152] But daily receipts soon declined "from the rapid increase in the number of cabs in use and perhaps also from a slackening of interest in the new conveyance as the novelty wore off." The incentive system had the further "grave defect" of encouraging speeding, thereby increasing the risk of accidents and decreasing the service life of the vehicle. Hungry drivers were concerned only that their vehicle survive until the end of their shift, prompting companies to experiment with various plans to encourage responsible driving and minimize repair costs.[153] And in 1908 falling wages precipitated another strike, directed in this instance against the New York Taxicab Company.[154] Moreover, like the early electric cabs, the internal combustion vehicles were prone to mechanical failure, prompting several cab manufacturers to employ modular engine designs that could be exchanged much like a battery tray.[155] Despite such efforts, many of the gasoline cabs had poor service records and were removed from service soon after their introduction, and the speculative investment boom in urban taxicabs meant that profits were both thin and uncertain. In short, the gasoline cabs suffered from many of the same problems as had the electric cabs when they were first introduced. The internal combustion engine was no panacea, either for drivers or owners.

Finally, in 1910 Meade committed to standardizing the fleet using the DeDion-Bouton gasoline vehicle and prepared to liquidate the remaining rolling stock. In a special taxicab issue of *Horseless Age,* he explained his strategy: "with the exception of good management, there is probably no factor of greater importance in contributing to successful taxicab operation than a standardized equipment." Citing the early experience of the Electric Vehicle Company, Meade noted that "every form of conveyance" had been produced: "the embarrassment of riches was complete and its own undoing . . . half the equipment was always out of season." The decision to close down the electric cab service was not due to any technical failure of the electric vehicles per se but to the increasingly attractive capabilities of the gasoline taxicab. According to Meade, the Mark 67 Landaulet—the last EVC electric cab model and "the most serviceable electric carriage ever built"—was "unexcelled for motor cab purposes, but for the unquestioned superiority of the gasoline cab on account of its unrestricted radius of action and ability to use pneumatic tires."[156] The lighter weight and pneumatic tires also yielded important safety benefits: the gasoline cabs, despite their higher speed, were easier to control and resulted in fewer damage claims against the company.[157] By 1909 the gasoline cabs averaged approximately fifty miles per day, considerably more than could be expected from even the Exide-equipped electric vehicles without remote battery exchange. Although there were more bumps in the road ahead, by 1916 the internal combustion vehicle was finally capable of providing dependable cab service with greater radius of operation than was its electric counterpart.

ASSESSING THE ALTERNATIVE SYSTEM

To assess the claim that the EVC cab service represented the seed of an alternative system for motorized road transport, we must consider several additional findings and arguments. First, the EVC scheme was created within the context of New York's horse-based transportation system. Horses and carriages, even privately owned ones, were housed in livery stables that were geographically concentrated at some distance from residential districts. Few people, even among New York's wealthiest elite, chose to keep their horses on site; many hired transport service from liveries. Under these circumstances, housing vehicles and people under the same roof was neither an obvious nor a trivial step. *Electrical Engineer,* for instance, predicted that many "electric carriages will be 'up-kept' by central charging stations and called by telephone whenever needed by the owners" but also hinted that the carriage might "be kept right on the premises, just as the bicycle is now." The journal continued: "There are scores and scores of fairly well-to-do families that would welcome the chance to 'set up a carriage' at the cost of sacrificing part of the semi-useless basement."[158] In this respect, the electric vehicle service plan was consistent with existing practice and should be seen as a natural outgrowth of prevailing, horse-based norms, even as the spread of the motor vehicle offered the possibility of radically altering those norms. Customers simply hired an electric vehicle instead of a horse and carriage, producing an incremental and functionally transparent shift from horse to motor vehicle.

Second, coordination between road and rail transportation providers was crucial for the EVC's success. Although there was ample precedent for conflict between these groups, the model of integrated service articulated by the Whitney-Philadelphia interests was similar to the relationship between gas and electric utilities in which companies evolved to offer both gas and electric service.[159] In the case of the Metropolitan Street Railway, the significant interlocking holdings among the streetcar company, the cab company, and the battery manufacturer (ESB) underscored both the practical synergies and the spirit of cooperation that organizers initially hoped for. A September 1900 report to Thomas Edison titled "Automobiles versus Trolley Cars" offered several insights into these relationships. Prepared in the midst of the EVC venture, the study noted "little apprehension, if any" in "railway circles" about the "future supplanting of the modern street car by the more modern automobile." Specifically, despite the "fond hopes of many New Yorkers when the present automobile service [EVC] was established that they could be taken from their homes uptown to their offices for fifty cents," in practice it cost "not less than $2.50 . . . to take a passenger from 72nd Street to Wall Street."[160] Initial fears about the imminent collapse of demand for streetcar service were thus unwarranted; in the short run, the street-car lines were not likely to lose business to the motor vehicle. But the report went on to describe extensive experimentation "in the past two years in the matter of building omnibuses to compete directly with the street car." All the efforts had thus far resulted in failure, but "the electric vehicle had probably come nearer accomplishing the results than any other." The report concluded that, although a mo-tor bus would have little chance to outperform a streetcar in head-to-head competition (costs per passenger were too high based on the modest-sized vehicles that had been developed up to that time), the most appropriate role for the motor bus would be as "a feeder for the street railway rather than as a competitor." Within the electrical industry, the initial experience with the electric cabs was neither so positive that it created enmity between supporters of rail and road nor so negative that it soured the electrical interests on the ultimate value of electric-powered road transportation.

Industry observers also recognized the importance of the relation-ship between the established street rail interests and the new motor vehicle company. "The electromobile may be turned into a source of revenue to the railway in at least two different ways," noted *Electrical Review*. "Current may be sold to the owners of the vehicles, thus offsetting the small loss due to decreased patronage. In the second place, the electromobile may be operated as an adjunct of the street railway (lines of motor omnibuses being run beyond the termini to parks and the like), as connecting lines between separate railway lines, or even as feeders in city streets where rails are objectionable or track construction unlikely to prove remunerative."[161] Others pointed to the parallel between the electrification of horsecars in the 1890s and the expected electrification of horse-drawn road vehicles in the 1900s. By late 1898 *Electrical World* claimed that "the electric automobile in-dustry is in a stage comparable to that of electric traction about ten years ago—namely, just on the threshold of a tremendous expansion." By the time of the National Electric Light Association's annual meet-ing in New York the following spring, *Electrical Review* expressed what was fast becoming conventional wisdom: "It does not require a prophet to see that the future of self-propelled electric vehicles will be a great one, or to note the fact that the automobile is, today, where the electric street railway was ten years ago in point of development."[162] The vision of an integrated, electrically powered, urban transportation system combining rail and road-based vehicles was revolutionary but not ridiculous. Given recent experiences first with the initial spread of electricity in the 1880s and then with the electrification of horsecars in the 1890s, many within the industry firmly believed that electricity would continue its expansion into new markets. Few members of the electrical fraternity would have disagreed with *Western Electrician*'s conclusion that "electricity is the natural medium for the application of motive power. Its supply is unlimited. It is everywhere. It is to movement what the sun is to growth."[163] Taking electricity off the rails was a logical and almost inevitable step.

Finally, in assessing whether or not the EVC experiment was a full and fair test of the idea of electric-based transportation service, we must admit that the electric cabs and the street railways of New York never actually achieved integrated operations of the sort proposed in the aforementioned reports, articles, and editorials. Cab service remained an indulgence of the relatively well-to-do; the masses of New York commuters continued to depend on streetcars or their own two feet; and the electric bus—the proposed "adjunct of the street railway"—was never able to establish a lasting place within the transportation system. Despite repeated efforts, for instance, the Fifth Avenue Coach Company could not makes its electric buses pay.[164]

In the end, the street railways and the electric cabs were unable to create any sustaining link or (in contemporary parlance) synergy between their operations. The Metropolitan Street Railway Company, the New York Transportation Company, and the Fifth Avenue Coach Company continued to provide different services to different markets. Far from taking advantage of its pioneering role in introducing electric vehicle service, the cab company ended up paying a steep price for innovating. For too long, it was burdened with aging, nonstandard, horseless carriage–era equipment. Meanwhile, new entrants were introducing modern, automobile-based designs. Far from being helped by their close association with the street railway, Meade, the New York Transportation Company, and the Fifth Avenue Coach Company were hampered by the prevailing ownership structure. As late as 1920, Meade reported that the Fifth Avenue Coach Company "has never grown as fast as the public demand for its service warranted—primarily because of its control by the rail traction interests.

In other words, its phenomenal growth has been largely *in spite of* conservatism and restriction, and not *because of* enterprise or encouragement."[165]

Far from creating an opportunity for the future development of an electric vehicle–based urban transport system, the shadow of the EVC hung over the industry for years. Reviewing the early history of the electric cab venture, a 1908 column in *Power Wagon* observed that the "New York concern . . . had the misfortune to be handicapped at the outset of its career with a lot of bad investments. Its family relations with the manufacturing company which supplied its equipment did not tend to keep engineering up to date."[166] Regardless of its many achievements—the first electric vehicle service in the world; the development and construction of a series of innovative, mechanized central station designs; the successful operation of a fleet of six hundred electric cabs—the EVC is remembered as a failure. By the end of 1907, William C. Whitney had died. The Metropolitan Street Railway—handicapped by overcapitalization, restrictive fare covenants, increasing franchise taxes, an unworkable mélange of prime movers, and growing competition from the Interborough Rapid Transit subway—was bankrupt, a victim of the Panic of 1907. With the advent of the New York Public Service Commission and the expanding regulatory roles of municipal and state authorities in the provision of public services, the prospect of an integrated private transportation system combining electrically powered streetcars, buses, and passenger vehicles was all but extinguished. There would be no integrated electric transit system in New York or any other American city.

Technological limitations alone do not explain the ultimate collapse of the EVC. The early electric cabs were capable of providing satisfactory service, but they required careful, routine maintenance to perform at their best; and supervisors and drivers often neglected all maintenance until the vehicle stopped working completely. Often by this time it was too late to effect minor repairs, and whole batteries or controllers were rendered useless. The social technologies surrounding inspection and maintenance appear to have been at least as important for successful service as the physical characteristics of the components themselves.

The attacks leveled against the Electric Vehicle Company were accurate in many respects, but prior historians, by accepting the extreme claims made by E. P. Ingersoll in *Horseless Age,* have overlooked the EVC's many important accomplishments. The experience of the

New York Transportation Company demonstrated that specific types of electric vehicle service could be profitable. Following the successful introduction of the Exide battery, the electric cab service produced an operating surplus until the drivers' strike in the fall of 1906. Thereafter, Richard Meade found that gasoline cabs were more profitable to operate than was the aging electric fleet. His logic, however, was based on increasing daily mileage and profit sharing that rewarded drivers for traveling longer distances. The average cab mileage gradually outgrew the efficient range of even the Exide battery so that without a dense network of battery exchange stations the electric vehicles were at an increasing disadvantage. During the life of the cab fleet, population was spreading out; range requirements were gradually increasing; and the electric alternative continued to lose ground as investment, innovative effort, and public tastes were drawn to internal combustion. By contrast, had the electric vehicle system prospered, battery standards, exchange facilities, and other vehicle pooling schemes might have taken root and put internal combustion at the disadvantage, at least in cities. The Broadway cab station in New York City pioneered the principle of mechanized battery exchange that, if widely practiced and supported in other areas by regional central electric stations, might have established a pattern of for-hire electric transport service in major American cities. Although historians have passed a negative judgment on the EVC and found flaws in its vision for an electrified urban transportation system, the company might have succeeded. If it had, we might today mention Pope and Whitney in the same breath as Edison and Westinghouse: brilliant system builders who have transformed our daily lives.

Finally, it is clear that the EVC saga colored perceptions of the electric vehicle for many years following the collapse of the national scheme. This effect was, in some sense, inevitable. An expansive plan was necessary if the electric vehicle was to emerge as a serious alternative to internal combustion on a citywide or regional basis. Small-scale pilot projects, while perhaps useful to demonstrate, validate, and troubleshoot relevant technological components, would not have forestalled the arrival of internal combustion. Only a large-scale regional system, such as that envisioned by Whitney and Pope, would have provided sufficient coverage to preempt internal combustion. The EVC had to be big to stand a chance of succeeding, but its very size ultimately confirmed public perception of the venture as a massive failure.

CENTRAL STATIONS AND THE ELECTRIC VEHICLE INDUSTRY
Organizing to Support Electric Vehicles, 1900–1925

The time has arrived when traction batteries, under conditions of vehicular traffic, are making excellent records and showing commercial results that are most satisfactory. At last this element of greatest mystery and uncertainty, and the one that has beclouded the electric vehicle world from the first, has cleared from the horizon, leaving only known factors with which we may deal with certainty. The success of the electric-vehicle being thus assured, a steadily increasing demand for current is sure to follow, which will result in large business. This field belongs to the central station by right of discovery; but history repeats itself, and the opportunity will be lost if the territory is not occupied and its resources developed. The central station organization and equipment is the proper guardian and natural source of expert service and attendance for both battery and machinery, and will naturally be looked to foster the interests that it shares with vehicle operators.

> Elmer Sperry, in a paper read by George Sever before the twenty-third National Electric Light Association Convention, Chicago, May 23–24, 1900

The electric automobile is a thing of beauty, and could be made one of great utility. The motors and the batteries are perfect mechanisms, central stations are in need of the load, and the only thing that appears to be in the way of a remarkable taking up of the electric automobile idea is the lack of intelligent coordination of the various industries represented.

> Editorial, *Electrical Review*, February 13, 1909

The average central station, for off-peak power, would require no additional investment beyond the charging set, which is very moderate, and the average cost of such power certainly should be, if anything, less than one cent per kilowatt-hour. . . . I believe it is well worth while for the central station to give that matter careful consideration. Put your house in order and keep in view the getting control of this business while it is still in the very beginning.

> Charles P. Steinmetz, before the thirty-seventh National Electric Light Association Convention, Philadelphia, June 1–5, 1914

> The electric vehicle is without question one of the most desirable pieces of electrical utilization equipment for load building purposes from the standpoint of the central station. It provides an off-peak load, improving the load factor. It is itself, as it travels through the streets, a constant advertisement of the use of electricity for transportation purposes. . . . The acceptance of these general fundamental facts should stimulate the representatives of the central stations to greater effort to extend the use of the electric vehicle even though this may require specially trained sales forces and special methods to overcome the close competition from other means of transportation.
>
> John W. Lieb, report before l'Union Internationale des Producteurs et Distributeurs d'Energie Electrique, Paris, July 5–10, 1928

For nearly thirty years, leading figures in the electrical industry touted the merits of the electric vehicle and its special value to the owners and operators of central stations. For his part, John Lieb had much to be proud of as his professional career drew to a close. In 1882, it was Lieb who, as a young apprentice, midwifed the era of electric lighting and distribution when he started the generator at the first Edison central station on Pearl Street.[1] Over four and a half decades, he held nearly every position in the electric utility industry. He rose through the ranks at New York Edison; was elected president of the powerful National Electric Light Association (NELA); and served during the First World War as chair of the National Committee on Gas and Electric Service, an arm of the Advisory Commission of the Council of National Defense. Of all people, Lieb was in a position to mobilize the central station industry in support of the electric vehicle. Yet he failed to produce many converts. Within the electric utility industry, few seemed to share his abiding enthusiasm. Although the industry mounted a series of campaigns to encourage the use of electric vehicles during the decade preceding American entry into World War I, by the end of the war prospects for the private electric passenger vehicle had all but evaporated. Modest support for the electric truck continued into the 1920s (see chapter 4); but that, too, gave way to the full-scale adoption of internal combustion. Not until the early 1960s did the electric utility industry rediscover the potential of the electric vehicle market.

By 1906 the gasoline-powered automobile was well on its way to becoming an important component of the American transportation system. The idea of the automobile had become widely known and accepted throughout the country, and the essential components of the modern automobile system were beginning to emerge. Given exponential growth in the production and use of gasoline passenger automobiles, what would be the fate of electric vehicles? Would they continue to share the road with gasoline cars, or would they slowly disappear into junkyards and museums?

Through the middle of the 1920s, groups of individuals and companies organized many different institutions to encourage the use of electric vehicles. Ranging from national manufacturers' associations to local car clubs to a special electric vehicle section of the NELA, these efforts took many forms and supported a wide array of activities. Electric vehicle manufacturers, electric central station companies, and storage battery makers all participated to a greater or lesser extent in these ventures. Managers of central stations were prodded to use electric vehicles for their own transportation needs, to offer reduced rates for charging vehicles during periods of slack demand, and to develop and operate public garage and charging facilities. Cooperative publicity campaigns were mounted to counter the negative image of the electric vehicle, to highlight its general advantages as an urban alternative to the horse, and to encourage motorization for commercial services. Local groups organized social events, outings, parades, and vehicle demonstrations. One committee sponsored research to develop a standard charging plug to allow vehicles of different makes to be charged at remote locations. All told, several hundred thousand dollars and countless hours of discussion were spent in support of the electric vehicle proposition—the belief that the electric vehicle could provide economic and efficient transportation service for most of the needs of the typical motor vehicle user.

We focus here on the genesis of these now-forgotten organizations and their relationship to the rest of the automotive and electrical industries. What was the impetus for the creation of the flagship Electric Vehicle Association of America (EVAA) in 1910? Why did electricity providers, vehicle manufacturers, and the battery industry join forces when they did and not earlier or later? What did the organizers hope to do, and how much did they in fact accomplish? Were these activities able to boost sales of electric vehicles, or were the publicity efforts completely ineffective in the face of the internal combustion juggernaut? Is the lesson of the EVAA one of liberation, of the power of the individual in the marketplace conquering vested interests? Or is it a story about the failure of the market to produce and sustain a range of potentially useful technological options, a story of too little variation and too much selection? In this chapter we examine the issues and debates that preoccupied the supporters of the electric vehicle within the context of the development of its parent sectors: the automobile, storage battery, and electric utility (central station) industries.[2]

THE ELECTRIC VEHICLE INDUSTRIES DURING THE "DARK AGE"

After the collapse of the Electric Vehicle Company and its promise of fleets of electric taxicabs in every major U.S. city, the market for stand-alone electric vehicles, both passenger and commercial, continued to expand but only at a modest rate. While demand for internal combustion vehicles grew spectacularly, the electric vehicle entered into what electric vehicle pioneer Robert McAllister Lloyd called a "dark age." Electric passenger vehicles were less prominent at the major automobile shows, and those vehicles that were available embodied few technical or design improvements. According to Michael Schiffer, the disappearance of the electric vehicle from public view "tacitly reinforced the perception . . . that the electric pleasure vehicle was no longer a factor in the industry."[3] Although this view was mitigated by the continuing spread of electric vehicles for commercial purposes, and despite the continued expansion of electric vehicle use in absolute terms, the period between 1902 and 1908 witnessed the unqualified explosion of interest in internal combustion and the relative decline of the electric vehicle.

The Storage Battery Industry

The turn of the century was a watershed period for manufacturers of storage batteries. Intellectual property disputes were resolved, industry structure consolidated, and the battery itself emerged as a dependable solution to a range of business problems. Before 1895 the American electrical industry took a dim view of the storage battery: "if an electrical engineer recommended the use of a storage battery, he was immediately put down as a 'crank.'" Or worse: "It used to be remarked that a man who acted as advocate for storage batteries was apt to economize truth." Scarcely three years later, however, total shipments of battery plates had increased tenfold, partially in response to demand from the nascent vehicle market.[4] In late 1894 the Electric Storage Battery Company had obtained all battery-related rights, patents, and licenses formerly controlled by Consolidated Electric Storage,

General Electric, Brush Electric, and Electric Launch and Navigation.[5] By the time of NELA's annual meeting in Cleveland the following spring, opponents and supporters of storage batteries were arguing about the economics of storage battery operation but not its feasibility. Both sides agreed that there were cases in which stationary storage batteries could be successfully applied.[6] Although demand for electricity varied throughout the day, producing the typical load curve, shutdown and startup costs were too high to justify actually "drawing the fires."[7] Instead, stations used batteries to store electricity produced during periods when supply exceeded demand. Depending on the scale of station operations, batteries were also used to smooth current output, boost output at remote substations, dampen peak usage, and carry the load during outages.

The storage battery bandwagon continued into the twentieth century. As total electricity production increased, so did the scale and number of stationary batteries in use. The battery imported by Boston Edison in 1894, for instance, briefly laid claim to being the largest storage battery in the world, but it was soon eclipsed by the 22,400–ampere unit installed by Chicago Edison in 1898.[8] Addressing an international engineering congress in St. Louis in 1904, L. B. Stillwell boasted that total installed storage battery capacity had increased a hundredfold during the preceding ten years.[9] Several months later, an annual review of the electrical industry observed that "an accumulator [battery] auxiliary is now recognized as an essential part of the equipment of every important generating station."[10] And by 1909 the use of batteries "as insurance against interruption of service" had become "so universal that it may be considered conventional."[11] To give us a sense of perspective, fifteen years after the introduction of stationary storage batteries into American central station practice, New York Edison alone operated forty-one batteries with total storage capacity of 193,000 amperes. By 1916, although the rate of growth had slowed, the number of batteries in use had expanded to forty-eight, with total capacity of 225,000 amperes. By this time, however, practice had shifted away from load smoothing toward shorter and heavier emergency discharges.[12] Notwithstanding changes in how stationary batteries were used, the general trend was clear: like the automobile industry, the storage battery industry experienced dramatic growth during the two decades preceding World War I. In the absence of radical technological change, observers attributed the "recent high favor of storage plants" not to "the increased output of storage cells when first charged and discharged, but rather to improvements in manufacture, whereby the depreciation has been reduced to commercial limits."[13] Improved durability and extended battery life had established an important market for the storage battery in the operation of the central station.

The Central Station Industry

The electric utility industry—in particular, a handful of leading central stations—played an important role in rekindling public interest in the electric vehicle. Like the gasoline automobile sector and the storage battery industry, the electric utility industry experienced a period of dramatic expansion during the first decade of the century. Using Chicago Edison (later Commonwealth Edison Company) under the leadership of Samuel Insull as an example, we find that the tumultuous 1890s gave way to a long decade of growth and prosperity. During the 1890s Insull, propelled by the vision of unified electric light, traction, and power distribution, embarked on a campaign to consolidate power production by acquiring small, local generating stations. He paid a financial premium to create a regional electric monopoly; from 1892 to 1898 Chicago Edison paid nearly twice as much for its acquisitions as the tangible book value indicated they were worth.[14] With much of its overcapitalized system still based on older, direct-current generation and distribution equipment, the company's efforts to increase electricity consumption were vulnerable to threats by large customers to disconnect from the central station lines and build their own isolated generating facilities. To forestall the loss of these large consumers, Insull introduced two-tier, preferential rate schedules based on apportioning use between peak load and total consumption. This strategy allowed him both to hold onto his growing industrial power market and to expand coverage and service throughout the greater Chicago region. As a result, he was able to diversify the electric distribution load and surmount the initial threats of overcapitalization and the spread of isolated plants.[15]

After experiencing rapid growth and sluggish profits during the 1890s, Insull reaped handsome rewards in the following decade. The introduction of the high-speed turbine generator—first installed at commercial scale at the Fisk Street Station in Chicago in 1903—enabled him to continue the process of building the load. Whereas earlier consolidations were driven primarily by the logic of load diversity (that is, the more diverse the load base, the less likely that users would demand electricity at the same time), the turbine generator, in conjunction with high-power alternating current distribution, actually lowered the cost of producing large quantities of electricity and allowed Insull to reduce prices and still make profits.[16] The number of isolated plants installed annually in the Chicago area declined from 121 in 1900 to 17 in 1910.[17] The expansion of the Edison system also allowed Insull to tap into the power market for local transit. From 1902 to 1907 Chicago Edison increased its share of transit power from 1.6 to 29.7 percent, while gross electricity production for transit increased more than fiftyfold (from 2.6 to 148.8 million kilowatt-hours).[18] Moreover, a series of well-orchestrated sales campaigns introduced consumers to a host of new electric technologies.[19] As a result, the Edison load curve, measuring average use as a percentage of peak use, rose from 30.4 percent in 1898–99 to 41.7 percent in 1910–11.[20] Having combined these managerial strategies with deft political maneuvering, Insull emerged by 1909 as the leading prophet of the gospel of consumption. His strategies were institutionalized, and his slogan—"low rates mean good business"—spread throughout the industry.[21]

Despite the larger trend toward load building and marketing new applications of electric power, the electric vehicle market did not generate positive results for central station balance sheets until at least 1907. Previously, the automobile charging market had been disappointing. As early as 1899, observers had encouraged central station managers to demonstrate and promote the importance of the electric vehicle load: only "the example of the whole electrical fraternity itself going in for [electric] automobiles" would provide the stimulus needed to establish a sustainable market for charging current.[22] Soon the same observers were criticizing the central stations for their apathy. Comparing the central stations' response to the electric vehicle to their reaction to the electric power motor, *Electrical World* recalled that the power motor "had literally to be choked down the throats of many central station managers" and expressed hope that the same managers would be more open minded this time around.[23] Yet the contribution of the electric vehicle to the overall central station load curve was less desirable than expected. Despite high hopes for the sale of current to electric vehicle users, inventor Elmer Sperry acknowledged in 1900 that "the demand for current is not as favorably located on the load curve or on the network as was hoped." Because most commercial charging varied seasonally with other demands for

electricity and because there was little effort made to educate and attract vehicle owners to the benefits of off-peak charging, the electric
vehicle load failed to increase the diversity factor. Sperry's proposed
solution was to use "duplicate traction batteries" to enable charging
to "go forward at times and rates dictated by the station and practically under its control." In addition, Sperry advocated multiple-rate
meters to encourage garage operators to take advantage of cheap and
plentiful off-peak power.[24] Regardless of these early setbacks, he remained optimistic about the future market. Sperry himself had developed several valuable patents in the field of battery design, which
he later sold to the American Bicycle Company, and he believed that
between the limited availability of gasoline and the improved performance of electric vehicles the chances were good that central stations would eventually profit from the sale of electricity for vehicle
charging.[25]

For a minority of central station managers, the rapid spread of
the automobile could not help but lead to an expanding opportunity for vehicle charging. Responding to the sense that the "advent
of the automobile" had "entirely escaped the notice of central station managers," A. E. Ridley, in a 1903 talk, outlined three important attributes of the electric vehicle load. First, unlike individual
lights, irons, fans, and hot plates, the electric vehicle consumed a large
quantity of electricity. Second, electric vehicles could be charged
"when the station has plenty of power which central station managers are unable to use." Third, the "character of the load is absolutely
ideal." Unlike other applications and appliances, which were turned
on and off frequently, charging a vehicle storage battery was "perfectly regular" and well known to those managers, who were accustomed to using stationary storage batteries for other purposes. In
Ridley's view, the central stations were missing a valuable opportunity: "if the companies would install motor converters and operate
wagons of their own, their example would be followed by many who
are now deterred by the inability to secure direct current suitable for
charging their machines."[26] Several lighting companies were already
following his advice. By early 1904 Boston Edison owned four electric vehicles—two trucks and two superintendents' cars—for its own
transportation requirements and had installed twelve motor generators to allow vehicle charging at various locations within its service

area. These locations, along with possible touring routes, were advertised in *Edison Light,* a monthly newsletter sent to all customers. W. H. Atkins, the general superintendent at Boston Edison, was a strong supporter of the electric vehicle, believing that the associated charging facilities not only provided convenient charging for electric automobile owners but also generated income for the central station and were "an excellent and permanent advertisement for the company."[27]

Other central stations also experimented with electric vehicles. In 1902 New York Edison, for instance, had taken delivery of its first electric truck, an experimental five-ton vehicle made by the Electric Vehicle Company that was equipped with motors for each wheel and a separate steering motor intended to make the vehicle more maneuverable.[28] A year later the company had purchased at least six more electric vehicles for the use of its regional superintendents.[29] The vehicles performed adequately for several years, and the fleet gradually expanded, numbering fifty-three by early 1906.[30] It is thus significant that even in the depths of the dark age of electric vehicles, some electric utility operators were groping toward an effective strategy to use vehicle charging to build the electric load. It is also clear, however, that Boston Edison and New York Edison were exceptions to the norm.[31]

During this period of transition, profits from the sale of current and the underlying economics of motor vehicle operation were less important than boosting the public appeal of modern electric service. Among leading central stations, sales campaigns and the techniques of selling were becoming more sophisticated. Edison affiliates in Boston, New York, Brooklyn, Chicago, and Philadelphia had each begun to publish monthly bulletins highlighting the spread of electricity.[32] The use of electric vehicles (often illuminated with electric lights and decorated with signs proclaiming the benefits of central station service) to deliver bulbs and light street lamps kept electricity in the public eye. And because the central station did not pay to use its own mains or draw on in-house technical expertise, the real costs of electric vehicles compared with the costs of the horse-drawn carts they replaced or those of prospective gasoline service were unknown and, for the time being, immaterial. It was sufficient that the electric vehicles answered the calls and lit the lamps.[33]

HAYDEN EAMES, THE "APOSTLE OF THE JUICE WAGON," AND THE EMERGENCE OF ORGANIZATIONS SUPPORTING THE ELECTRIC VEHICLE

The growth of the gasoline passenger automobile industry; the success of the stationary storage battery; and the convergence of concerns about advertising, load building, and economic viability set the stage for the electric vehicle industry to grow beyond the limited role established by the leading central stations. In late 1906 electric vehicle producers—including Studebaker, Baker, Pope, and General Vehicle—established the Association of Electric Vehicle Manufacturers to advance the industry.[34] Several months later, Hayden Eames, a former naval officer who, more than a decade earlier, had recruited Hiram Percy Maxim to join the Pope Manufacturing Company, addressed an assembly of the new organization. Warmly welcomed as the "apostle of the juice wagon," he regaled his audience with tales of the founding of the electric vehicle industry in Hartford before turning his attention to the prevailing state of the market. Citing examples of the Adams Express Company in Buffalo and the New York Transportation Company in New York City, he reported that, despite the persistent "chemical peculiarities" of the storage battery and other common misperceptions, the electric vehicle was succeeding in providing valuable transportation service. Interest among potential users was high. "Everywhere we go," said Eames, "we always find them interested, but still there is doubt as to just *what* to do."[35]

Later that year, Herbert H. Rice, a manufacturer of Waverly electric vehicles, laid out the electric vehicle agenda in an address to the NELA on the renewed opportunity to sell current for vehicle charging. Not since Sperry's talk seven years earlier had the national delegates been challenged to seek out the vehicle charging load. But while Sperry had been content to draw attention to the future prospects for electric charging, Rice was more outspoken: "Central-station men, who should have been foremost in advocating and pushing the sale of electric automobiles, have been most apathetic." Attributing this neglect in part to the "lamentable financial failure of early endeavors to utilize electric vehicles in cab service [that is, the EVC]," he also blamed it on station operators' continued ignorance and their misguided belief that other sales opportunities were more lucrative. Even with their limited range and speed, electric vehicles could go "much faster than the city laws allow" and cover "sufficient mileage to satisfy all requirements of city use."[36] Moreover, central stations were overlooking electric vehicles at the same time they were promoting other appliances with less attractive load characteristics. For Rice the solutions were obvious. First, central stations should "quote favorable rates and encourage the sale of current." Second, every central station should appoint a local representative to "take a little time to find out what the carriage owner needs in order to take current from your mains." Third, central stations should purchase electric vehicles for their own transportation needs: "What will you think of an electric light company that burns gas for its own lighting? How about those whose officers and managers who use horses instead of electric runabouts and employ horse-drawn wagons instead of electric trucks?" "Work for work," Rice concluded, the electric was "better and cheaper."[37] Thus, by 1907 the stakes had been raised. Electric vehicle makers saw the market for their products beginning to rebound and were irate about the lack of support they perceived from the central stations that stood to gain from the spread of their product.

Less than two years later, on March 11, 1909, a score of business leaders gathered in the Boston Edison offices to establish the Electric Vehicle and Central Station Association (EVCSA). Electric Storage Battery Company representative Frank J. Stone took responsibility for prompting the community into action, claiming that a presentation by "a prominent storage battery manufacturing company" in November 1908 had "excited a great deal of interest" and was "the direct cause of a number of central stations advocating the use of the electric [vehicle]."[38] Among those present were central station executives, electric vehicle manufacturers, and representatives from the storage battery industry. Together, they spearheaded a range of activities in support of the diffusion of electric vehicles. Their motto—"to encourage the adoption and use of electric commercial and pleasure vehicles by electric light and power stations and their customers"—showed that the association initially saw the central station industry as its primary focus for action.[39]

The stage seemed set for the vehicle branch of the electrical industry to catch up with the accomplishments of its related sectors. The vehicle makers were organized; the central stations, battery manufacturers, and electrical suppliers were "tied together by bonds of commercial enterprise." Observers saw no reason why they could not "get together and establish some comprehensive scheme where arrangements could be made for charging stations in at least all of the large cities and in most of the small cities where electricity is now used in any form."[40] Believing that there had "been need of such an organization for a long time," editorialists at *Electrical Review* predicted that "the new association should meet with immediate success."[41] Summing up, the journal proclaimed that the electric automobile was "a thing of beauty, and could be made one of great utility. The motors and the batteries are perfect mechanisms, central stations are in need of the load, and the only thing that appears to be in the way of a remarkable taking up of the electric automobile idea is the lack of intelligent coordination of the various industries represented."[42]

As we can see from descriptions in *Central Station,* the early meetings of the EVCSA were devoted mostly to building awareness of the electric vehicle. Participants read papers purporting to demonstrate the perfection of the new electric vehicles, reported recent sales, and raised glasses in toasts.[43] Riding the rising tide of enthusiasm, the principals of the EVCSA realized that their regional organization would not be able to mount an effective national movement in support of electric road transportation on its own. Accordingly, a Plan and Scope Committee led by President William H. Blood was appointed to explore the possibility of creating a national organization modeled on the National Electric Light Association. At a June 8, 1910, meeting attended by Thomas Edison himself, the general plan for the creation of the Electric Vehicle Association of America (EVAA) was approved. Formal arrangements crystallized over the summer, and during the New York Electrical Show in October 1910 the Electric Vehicle Association of America was introduced to the public.[44]

THE ELECTRIC VEHICLE PROPOSITION: THE SPREAD OF AN IDEA

Following the creation of the association in 1910, spirits were high, even as the membership recognized the magnitude of the challenge they faced. Writing in *Central Station,* Edward S. Mansfield, a senior Boston Edison executive, indulged in "a little burlesque technicality" to describe the "electric vehicle proposition": "Electric Vehicle + Battery + Central Station + Consumer = Success." Success depended, in this view, on the positive contribution of each member of the equation: "Place the negative sign before any one . . . or reduce the efficiency of one single member, and the term on the right of the equality sign is at once changed to 'failure.'" No single component technology was capable of achieving success independent of the others. Mansfield acknowledged that creating an electric-based transport system was a "massive proposition" requiring "hard work . . . to open the door to the opportunity now knocking without." The rewards for success, however, would be commensurate with the effort demanded; Thomas Edison himself had only recently claimed that, "in fifteen years, more electricity will be sold for electric vehicles than for lighting." Mansfield went on to catalog a host of specific barriers standing in the way of success: poor charging facilities, lack of standardized charging plugs, insufficient data about the real operating costs of different types of vehicles, and inadequate attention to the needs and desires of prospective consumers. Although focusing almost exclusively on the problems accompanying the spread of the commercial vehicle, he nonetheless accurately characterized the prevailing liabilities of the electric vehicle system for both private and commercial users. Only an effective, coordinated electric vehicle system would result in the success of each of the component members.[45]

THE VANGUARD, THE FOLLOWERS, AND THE SILENT MAJORITY

When the EVAA began its national effort, the electric utilities of the nation were poorly coordinated, and a wide range of practices characterized reactions to the electric vehicle proposition.

The Vanguard

Boston Edison was the first central station to actively invest in marketing the electric vehicle. In 1901 it was among the first electric utilities to acquire control of all the competing power companies within its service area. According to historian David Sicilia, Boston Edison increased its investment in marketing its electric services during the first decade of the century. As we have noted, the company was one of the first central stations to recognize the potential benefits of stationary batteries and had already supported the electric vehicle by creating an internal electric vehicle department and sponsoring the Electric Vehicle and Central Station Association. But by virtue of its secure market position, Boston Edison was able to do more. In February 1911 President Charles Edgar, who claimed that his own electric vehicle had run more than 40,000 miles over seven years and had "never been laid up on the road," announced a $100,000 campaign to educate the public and to purchase electric vehicles for use by Boston Edison. In a telegram detailing the campaign, Edgar claimed that the utility had recently purchased thirty-four new electric vehicles and that it would purchase additional vehicles to replace 120 horses and 39 gasoline cars. Two weeks later, more than one hundred industry representatives were treated to a grand dinner to launch the campaign. A state-of-the-art charging and storage garage was opened on Atlantic Avenue in April; and in May a small team of researchers at the Massachusetts Institute of Technology was granted funds to study the relative economics of gasoline, electric, and horse-drawn delivery service.[46] Thus, by the time of the 1911 EVAA annual convention, Boston Edison had clearly established its leading position in support of the use of electric vehicles. In a talk before a group of electric vehicle enthusiasts in Boston in 1913, William C. Anderson, successful manufacturer of the Detroit Electric brand, graciously thanked Boston Edison for its assistance: "management has been broadminded enough to see that the advent of the electric vehicle in large quantities in this city would effectively solve one of the most vexing problems which the central station has to meet, namely, *a constant demand for their off-peak load*. . . . The Boston Edison Company have set the pace by purchasing and operating 56 electrically propelled vehicles. . . . [they] have put their stamp of approval on the Modern Electric Vehicle."[47]

Boston Edison's commitment to the electric vehicle encouraged others to follow. Ways of using electric vehicles to advertise electricity and to sell electric current had been proposed for many years. But even the most persuasive conference paper, upbeat editorial, or enthusiastic monthly meeting of an electric vehicle booster club could not create physical networks of charging stations, vehicles, and consumers. Only a sustained, concerted effort by the various participants could actually advance the prospects for electric vehicles. For its part, Boston Edison had demonstrated its willingness to support the electric vehicle proposition.

The Followers

Central stations in a handful of other cities—Chicago, Cleveland, Denver, Hartford, New York, Newark, Philadelphia, Rochester, and St. Louis—soon joined the electric vehicle bandwagon. Although these cities bore little physical or demographic resemblance to each other, they shared certain similarities in their approaches to the electric vehicle business. Central stations established electric vehicle or transportation departments to provide technical support and maintenance for electric vehicle owners, subsidized local service stations offering charging facilities for electrics, and, if possible, employed electric vehicles to meet their own delivery and service requirements. They also sponsored local EVAA chapters and helped local customers share information and experience, gave vehicle and garage owners the best rates on charging current, and devoted portions of their local advertising budgets to promoting electric vehicles.

No two companies adopted identical strategies. New York Edison, for instance, had already established an automobile station under the direction of H. T. Cameron in 1909. By 1910 the company was operating as many as seventy electric vehicles, ranging from 1,000–pound wagons to five-ton trucks. Overall, New York City led the nation in the absolute number of electric commercial vehicles in use and in their relative share of the commercial vehicle market. The company maintained a card file on each electric vehicle owner in the utility's service area, and Edison representatives periodically inquired to make sure that service was satisfactory.[48] Meanwhile, in Chicago, where the electric commercial vehicle had yet to take hold but the electric passenger car was still relatively popular, Chicago Edison was slower to adopt electric vehicles for its own fleet but more aggressive in its effort to support local garage owners.[49] Differences were even more apparent between stations in smaller cities. In Rochester, the Light and Railway Company started early and had a vehicle department employing eleven people by 1909. The company operated eleven electric trucks and eight electric runabouts and also became the agent for Exide batteries, General Vehicle Company wagons, and Diamond rubber tires.[50] By contrast, the Denver Gas and Electric Light Company did not establish its automobile bureau until 1910, when it purchased four electric vehicles for its own use. By 1913 the station had expanded to twenty-three vehicles, with more than eight hundred private passenger electrics in use in the metropolitan area. The company did not choose to represent any specific line of vehicles, but it did hire a battery expert who supplied "complete operating instructions to all owners of pleasure vehicles" and even supervised construction of batteries for use in the company's own trucks.[51] In St. Louis the Union Electric Light and Power Company began its work in 1907, establishing an agency for two lines of vehicles as well as an automobile department. Vehicle registration grew from 11 in 1906 to 17 in 1907 to 350 in 1910 to more than 700 by the end of 1912. The company then decided that the market was sufficiently competitive to "withdraw from active competition in selling" and announced its intention to give up its agencies and close its garage to the public.[52] Not every follower station adopted all of the available marketing techniques; however, their generally supportive stance toward the electric vehicle contrasted sharply with that of the majority of central stations.

The Silent Majority

In May 1909 *Electrical World* conducted a national survey of approximately 4,000 central stations requesting general information about the development of the electric vehicle in their respective service areas. Summarizing the findings, even the ever-optimistic journal writers were forced to admit "the tremendous inertia . . . to be overcome before any great advance or improvement can take place." Most of the stations failed to respond, and the responses of those that did indicated no interest whatsoever in the electric vehicle: "a common and characteristic response" was to write "Nothing doing" across the top of the survey "without any attempt to explain the reasons or justify the absence of electric automobiles from the given territory." Only twenty-six companies, fewer than 1 percent of those surveyed, reported using electric vehicles for their own transportation needs. Among these respondents, the five largest operators accounted for more than 75 percent of the electric vehicles in central station service, and the top ten cities accounted for 87 percent of the total. Fourteen central stations reported owning only one electric vehicle. Even allowing for modest reporting bias and the claim that surveys were "still straggling in," we can see that as of 1909 only a handful of central stations across the country were paying attention to the electric vehicle, either as a market for electricity or as a tool for advertising and delivering electrical service.[53] Supporters of the electric vehicle had their work cut out for them.

THE WORK OF THE EVAA

Like many industry associations, the EVAA operated through a network of separate committees. Some were short-lived entities established to undertake specific projects (for example, the convention committee), whereas others (such as committees on membership, publicity, standardization, insurance, rates, and charging stations) were continued from year to year.

Publicity: *From Education to Advertising to Sales*

As a core institutional activity, the EVAA sponsored an ambitious national campaign to introduce potential customers to the idea of using electric vehicles. At the same time, sales drives by individual central stations offered customers a range of incentives to experiment with electric vehicle service. Within the association, these campaigns exposed conflicts about the identity of intended customers and the task of selling an electric vehicle. Because EVAA participants were unable to agree on the desired outcome, the potential influence of national and local initiatives was dissipated.

As we have discussed, the utility industry had accepted the idea of selling electric power by the time Boston Edison and later the EVAA launched their electric vehicle sales efforts. But whereas other campaigns were able to clearly separate the private market for home appliances from the commercial market for electric motors, the electric vehicle campaigns were poorly focused and were therefore unable to maintain this distinction in the public eye. Private customers were willing to pay for qualitative improvements in performance and style, even if these cost more; they were not, however, willing to endure inferior performance at any price. Accordingly, "Social Distinction," a multicolor brochure promising unparalleled quality and refinement, introduced the electric vehicle to potential private customers in the Boston area.[54] Commercial buyers, by contrast, focused primarily on cost-effective transport. To reach these business customers, the EVAA placed advertisements in various trade (*Commercial Vehicle* and *Power Wagon*) and trucking (*Team Owner's Review, Dry Goods Economist,* and *Cement World*) journals.[55] But given that electric vehicles were widely perceived as fashionable ladies' cars and that EVAA members were simultaneously advertising the social distinction of the electric passenger vehicle, commercial customers were likely to be cautious. Some of these buyers eventually overcame their inhibitions and adopted electric vehicles, but in general the advertising campaign sent mixed signals to prospective purchasers. Although the private passenger car and the commercial vehicle were supposed to occupy separate realms, the national advertising campaign was unable to maintain that separation within its own organizational and marketing effort.

Advertising alone would not have resulted in the successful revival of interest in the electric vehicle. Sales effort was also required, and within the EVAA there was considerable disagreement about how to actually sell vehicles. For experienced passenger vehicle manufacturers such as Herbert H. Rice, selling a car consisted of offering the prospective purchaser a desirable product that not only satisfied stated functional needs but also appealed to personal tastes: "Most electric vehicles are bought by the people, not because of their mechanical efficiency or design of construction, but because of the lace on the door and the paint on the outside."[56] Marketing and advertising shaped preferences, but selling a vehicle ultimately depended on a sales agent's ability to unite an appealing product with a willing purchaser. There were abuses, to be sure. Some truck makers sold their vehicles "on the same basis that they marketed their pleasure vehicles" and were willing to promise too much to make a sale, prompting one manufacturer to complain that his colleagues would "sell a truck to climb a tree if a man wants it to do that."[57] The standard, product-based model of selling in the passenger automobile industry carried over into the commercial and electric vehicle markets to the disadvantage of the electric vehicle.

In contrast, the electric supply industry, by virtue of the lasting ties established between customer and central station, evolved a service-oriented approach to sales, one likely to appeal more to commercial customers. Stability, support, and service were integral components of the central station system: "What you are selling is service and that is what is lacking in a great many trucks. It is not the wheels, the shafts, or the engine, but it is the service which they will perform, and when you have received a customer's money for a truck your interest has not ceased. It has only commenced. It is for your interest to see that he gets returns."[58] To this end, many central stations offered electric vehicle customers various services and incentives. Before switching to its successful battery exchange plan, the Hartford Electric Light Company provided its customers with free charging current and service for six months; New York Edison offered free boosting charges.[59] William P. Kennedy, citing leading companies such as General Electric and Westinghouse, claimed that "it is the organization which makes the sales," not the individual sales agent.[60] In Kennedy's view, the sales agent or "commercial engineer" who represented the larger organization at the point of sale should be knowledgeable enough to help customers assess their transportation needs and even willing to recommend nonelectric alternatives should those prove better suited to the specific service requirements.

Not all central station executives agreed with Kennedy. EVAA member M. Robertson accepted that "we should not try to sell an electric automobile where a gas car should be sold" but worried that "we are entering on very dangerous ground when we instill that feeling into any salesman or electric car interest." His proposed solution: pursuing "the course that the electric automobile can be put in, except under very few conditions." After all, he mused, "how many gas car people do we know that would say . . . 'This is an electric car proposition'?"[61] Other senior association members also felt that electric vehicle sales representatives should not be ashamed to know more than their customers. At an EVAA meeting in 1912, attendees debated a proposal to establish a standard maximum speed for electric vehicles. Speaking against the proposal, one member voiced a heretical notion: "If there is that demand, why not supply a high-speed car?" H. Lloyd, a director of the Electric Storage Battery Company, responded sharply: "Some of the speakers here have stated that we have got to meet the desires of our purchasers. I think in this as in a great many other things, it is best to educate the public as to what is best for them and not always to give them what they want."[62] Lloyd, speaking for many of his colleagues, believed he knew what customers should want and was therefore opposed to giving in to customer desires. But in practice, as Robertson all but conceded, "educating" the customer bordered on misrepresentation. Tempted into promising too much, electric vehicle salespeople too often fell into the patterns established in the gasoline vehicle domain.

The EVAA's advertising and marketing effort was beset by internal conflicts between the two main approaches to selling. The automobile sales techniques produced marvelous results for the gasoline manufacturers. "Faster and cheaper" indeed sold better. In the commercial realm, each faction grasped part of the key to effectiveness: the vehicle producers recognized the importance of providing customers with a desirable product, and the central stations identified the need for continuing contact between buyers and sellers of electric transport service. In late 1914, *Journal of Commerce* editor Ellis L. Howland captured the problems inhering in these two approaches. After observing that the first automobiles "stamped speed and luxury and sport and extravagance as the ruling factors . . . and brought into the commercial side . . . a type of salesman with selling ideas totally unfitted for the latter-day conditions," he noted that "bad habits and wrong conceptions from the early days of the automobile" had persisted. Whereas "the gas car salesman tried to sell his customers *automobiles* and the electric car salesman electrical *machinery*," both

ultimately failed to offer what the commercial customer needed—
"practical, economical, understandable and efficient delivery utility."[63]

Finally, despite the EVAA's investments in education, advertising, and publicity, sales to new customers rarely exceeded more than
30 percent of total sales. Manufacturers touted the high number of
reorders as a sign that existing users of electric trucks were satisfied
with their purchases, but these figures also reveal the ultimate inability of the EVAA's collective publicity campaigns to attract new businesses to the electric vehicle.[64] Although gasoline cars practically sold
themselves, electric vehicles had to be sold without being oversold.
A 1915 speech by Baker Motor Vehicle Company representative
George Kelly reported that selling practices had, in fact, resulted in
many dissatisfied customers. "Oversold and badly sold trucks," he
claimed, "had done more to hurt the [electric] motor-truck industry
than any other one thing."[65] Kelly recommended that dealers needed
to be better trained to avoid such problems; but as we have seen,
changing behavior was easier said than done, underscoring the fact
that selling electric vehicles was a delicate and uncertain task. The
EVAA had no choice but to navigate among the different conceptions
of selling—selling enough without overpromising.

The Committee on Standardization

Another EVAA priority was entrusted to the committee on standardization.[66] As early as 1900, various interests had recognized the importance of standardization to the fate of the electric vehicle. Initially,
these efforts were not aimed at facilitating battery exchange but at
increasing the general viability of the vehicle market. Writing in *Western Electrician,* James Pumpelly, for instance, hoped that establishing
fixed numbers of electrical cells and an agreed-on range of battery
tray sizes would reduce uncertainty in the battery market and allow
manufacturers to increase batch sizes and lower battery prices.[67] By
contrast, *Electrical World* promoted standardization to allow vehicles
made by different manufacturers to share a common charging infrastructure. Drawing an analogy with the example of line voltage, the
journal argued that just as "it is better for all that there should be a
110 voltage than 110 different voltages," so, too, it is "necessary that
battery boxes should be standardized, that charging plugs should be
'universal,' and that every charging plant . . . should be equally useful and accessible to every vehicle that comes along and needs
'juice.'"[68] Several months later, the journal attributed the apathy of

central stations in part to the absence of "greater unanimity as to methods of standardization," especially for charging.[69]

As of 1910, however, the situation had not improved. Many manufacturers kept making charging plugs of their own design, hoping to encourage repeat orders from their customers. At least eight different types of charging plugs were then in general use, nearly all of which, according to committee chairman Day Baker, had been "condemned by the fire underwriters of the large cities." Although it was not difficult for a trained electrician or a "skilled driver that we *sometimes* find" to rewire a nonstandard connection "with a couple of pieces of copper wire and a few moments' work," it was evident that for the electric vehicle to come into general use an appropriate standard plug was needed.[70] Baker, the Boston sales representative for the General Vehicle Company and a principal organizer of the Boston Electric Vehicle Club, had already contacted most of the vehicle manufacturers in hopes of enlisting their support for a standard charging plug.[71] E. R. Whitney, vice president of the Commercial Truck Company of America, was handed the task of negotiating differences of opinion among the various parties. Two years later, at the third EVAA convention, two concentric standard plug designs were adopted by the association membership: a small one for charging passenger vehicles and a large one for heavy-duty commercial applications.[72] These designs had received the approval of the National Board of Fire Underwriters and were subsequently accepted by the Society of Automobile Engineers, American Institute of Electrical Engineers, and the Electric Vehicle Committee of the Incorporated Municipal Electrical Association of England.[73]

Issues of standardization, however, extended well beyond the need for a universal plug. The standard plug design established basic compatibility among different brands of vehicles, but other standards were needed to increase the efficiency of the electric vehicle system.[74] For instance, traditional lead-acid and Edison iron-nickel batteries operated at different cell voltages. The number of cells in any given battery tray determined the voltage at which the battery would accept current. Because there was no standard charging voltage for all the competing designs, rheostats were needed to bring the line voltage down to that of the battery, resulting in excess current being dissipated as heat. Fixed battery voltages would, in turn, have permitted standardization of controllers, motors, and other subsystems, allowing owners to exchange or upgrade individual components as better ones became available. Efficiency concerns also prompted the New England section to ask the national organization in 1913 to convene a committee to establish standard dimensions for different classes of electric vehicle tires.[75] Engineer Robert M. Lloyd went even further, arguing that there existed a natural monopoly for commercial vehicle service. His plan called for the creation of a "monopoly in the manufacture of commercial vehicles"—like that emerging for utilities under the umbrella of state regulation—that would lead to lower initial costs and "more economical operation."[76] Although the EVAA committee did not issue explicit standards for battery and motor voltages, preferring to send representatives to the appropriate meetings of the Society of Automobile Engineers, by 1912 most manufacturers had settled upon the forty-four-cell lead-acid design and the sixty-cell Edison tray, both of which were well suited to drawing current from 110–volt lines.[77] The increasing dependability of line voltages and the gradual adoption of standard charging plugs created more opportunities for interchangeability and probably contributed indirectly to the stabilization of battery design.

Concerns about standardization were later extended to the larger characteristics of the electric vehicle system. For instance, storage and maintenance garages were one area thought to be amenable to standardization. With levels of service varying enormously, EVAA members hoped that if the association established standards for appropriate garage service, current and prospective owners would have greater faith in the electric distribution and support system. To this end, the committee on garages and rates prepared a placard with the EVAA insignia intended as a signal to vehicle owners that the garage met the association's minimal service conditions (like today's familiar AAA sign). As of 1915, the committee had approved fifty-two garages.[78] Another standardization plan, put forward by Alexander Churchward of the General Electric Company, aimed to set a standard maximum speed for all electric vehicles. Because power requirements increased rapidly above twenty miles per hour, the race to build faster electric vehicles (in the absence of other technical efficiency improvements) was forcing manufacturers to sacrifice all-important range: "And the real cause for this increase is not that the engineer of any one company has found some new battery, motor, or tire, but because the salesman finds it easier to dispose of a car which will go faster than that of his nearest competitor."[79] Churchward was actually trying to even the playing field among different vehicle manufacturers in order to increase electric vehicle sales throughout the industry; but as he recognized, the incentive for an individual manufacturer to ratchet up speeds was too great to pass up. Only at the technical level could companies agree on the benefits of standardization.

Cooperation: The Anatomy of Institutional Action

One of the most important debates within the EVAA revolved around the general problem of cooperation. Even Elmer Sperry, back in 1900, had commented on the problems that electric vehicle owners experienced: "prospective customers have in instances been embarrassed by want of cooperation, aid and enthusiasm justly expected from the management."[80] From the beginning, therefore, cooperation was the touchstone of the association's success; and ineffective cooperation was, in the end, the cause of its eventual collapse. The EVAA's policy of promoting "the adoption and use of electrical vehicles for business and pleasure purposes" depended on successful cooperation among all major participating industries. Yet almost from the outset, both central stations and vehicle manufacturers refused to believe that the other arm was doing its fair share. Two complaints were paramount: the vehicle manufacturers claimed that the central stations charged too much for electric current, and the central stations claimed that the vehicle manufacturers were hindering the expansion of the vehicle market by keeping vehicle prices too high. Both charges were partially true. It was certainly true, for instance, that electric passenger vehicles were more expensive than comparable internal combustion cars.[81] The electric vehicle manufacturers stressed the high-quality appointments and sound mechanical construction of their products, but ultimately the industry had no answer to the much more affordable, mass-produced Model T. Consulting engineer William Kennedy attributed high prices to the small scale of electric vehicle production: "very few [manufacturers] are turning out a thousand cars a year; that means that the price must necessarily be much higher than the price of a gasoline car when we consider that Ford is turning out cars at the rate of 75,000, and possibly 100,000 by next year [1913]."[82] Many EVAA members pleaded repeatedly for an inexpensive, mass-produced passenger electric but to no avail. Despite persistent rumors about a Ford-Edison collaboration, an inexpensive, mass-produced electric vehicle never arrived.[83]

Concerns about the cost of current were equally pervasive and intractable, although for entirely different reasons. J. W. Brown, a representative of the Couple-Gear Company, summed up the frustrations of the manufacturers when he said, "The only factor that seems to influence or hold back the purchase of the battery-driven commercial wagon is the non-desire on the part of the central station people to give reasonable rates for charging storage batteries."[84] Brown may have been correct, but he was preaching to the choir. The central station managers whose minds needed to be changed were not within earshot. The members and other industry representatives who attended the monthly and annual meetings of the EVAA had already accepted the religion of the electric vehicle and were hardly the ones gouging vehicle owners for current. At its height in 1915, the association had only 1,000 members; and with nearly 6,000 central stations in the United States at the time, it is clear that many station managers were still unmoved.

Cooperation among the EVAA membership also foundered on the issue of sharing information about the cost of vehicle operations. A significant part of the association's committee work was dedicated to standardizing reports of the costs of using electric vehicles. Committees examined prevailing charging rates, garage costs, and general fleet operations, gathering in the process a small mountain of general data (see chapter 4). But as the head of the committee on operating records observed in 1913: "There is considerable reticence on the part of the electric vehicle owners to give any data that could be considered as authentic and exact. They will not even endorse a set of figures which would be prepared by competent parties as to the probable cost of their specific installation."[85] At the annual convention the following year, Waverly representative Herbert Rice demonstrated this reticence, citing his company's policy of not discussing actual numbers.[86] An editorial in *Commercial Vehicle* attributed China's "long lethargy" to the existence of the Great Wall and criticized truck owners for maintaining "a China wall about their individual installations, fearing to permit others to examine their problems and profit by their experience."[87] With vehicle operators guarding their true costs, prospective purchasers viewed whatever data were presented with increased suspicion. William Kennedy's commercial engineer faced a difficult challenge: "if we take a specific case and use the figures in that case, very few people, even in the same line of business, will regard them as sufficient evidence. . . . John Smith in the dry goods business on this side of the street might object to the cost of Bill Brown's business on the other side of the street."[88] Although the EVAA hoped to establish sound baseline estimates of the relative costs of operating different types of vehicles in varying commercial settings, potential consumers never quite trusted the results. Announcing a major survey to accurately document operating costs, *Electrical World* admitted that "so much of half truth and half fiction has been circulated on the performance of electric vehicles that the whole industry has been hampered in its legitimate growth." Variations in reported operating costs served "to destroy rather than to create any confidence in them."[89] Numerous indirect instruments—from novel accounting schemes to pocket hydrometers to the highly touted Sangamo amp-hour meter—were proffered to help confirm vehicle operations and costs but failed to solve the crisis of confidence.[90] Ultimately, despite many written guidelines and common rules of thumb, a driver, even with detailed knowledge of initial conditions, could never be sure exactly how far an electric vehicle would be able to travel and how much it would cost to get there.

In their effort to expand the market share of the electric vehicle, members of the Electric Vehicle Association of America repeatedly disagreed about the question of agency: who was responsible for what? At various points in the life of the association, representatives from different institutional groups temporarily assumed the mantle of the heterogeneous engineer, trying to seamlessly integrate the social and technical components of the electric vehicle system into a smoothly functioning whole.[91] But in the end, clear lines of accountability failed to emerge. No one party to the electric vehicle proposition was able to forge a working system from the pieces available, and the ideal of cooperation unraveled amid mutual suspicion and finger pointing. Various actors, from companies to industry associations to leading executives, tried to force cooperation on the participants, but for various reasons no faction was able to succeed. The first leaders of the reborn electric vehicle movement were the electric vehicle manufacturers themselves. Waverly, Baker, the Vehicle Equipment Company (later General Vehicle), and even the scaled-down Columbia line from the Electric Vehicle Company all struggled to establish a beachhead in the motor vehicle market. Up to 1906 these companies sold modest numbers of passenger electrics to wealthy urban women and perhaps 1,000 commercial wagons to central stations and select merchants in major cities. Before the creation of the Association of Electric Vehicle Manufacturers in 1906, however, the vehicle manufacturers were unable to generate sustained interest within the utility industry. The central station managers viewed the private electric owner as a weekend nuisance and charged for current accordingly. The few fleet operators using electrics before 1909 were often less concerned with economy than with advertising and the appearance of modernity. And as for Herbert Rice's plea that the central stations should at least use electric vehicles for their own transportation needs, most executives were not involved in the daily operations of their own transport service. They would no more ask the stable hands about present and future costs of horses or motor cars than they would help their wives and domestics rearrange the kitchen appliances. Class and status concerns relegated the stables to the lowest possible priority.[92] Charging high prices for current meant more revenue from a trivial and sometimes annoying set of customers. In response to cries from customers and manufacturers that monthly bills were too high, executives could easily point to deficiencies in the battery or charging technique. On their own, the vehicle manufacturers were unable to orchestrate the spread of the electric vehicle.

As the central station industry gradually warmed to the idea of the electric vehicle, station managers and the EVAA assumed principal responsibility for the success of the electric vehicle campaign. At the first EVAA convention in 1910, several speakers admitted that the central stations should do more to share the initial costs of popularizing the electric vehicle. Commonwealth Edison executive Louis Ferguson argued that the "central station man . . . cannot shirk his responsibility and must in the end leave a satisfied customer." "Overcharging [of batteries]," he said, is not the "cause of all battery troubles, but it is usually the cause of excessive electricity bills," and excessive bills were cause for concern, not celebration.[93] At the same meeting, consulting engineer William Kennedy observed that "almost the entire burden for the launching of a project of such scientific and practical magnitude has . . . rested upon the shoulders of the manufacturer." The creation of the EVAA was, in part, the result of the "long and persistent effort to *awaken the central stations* to the realization of the very *important* and *profitable share* which they should take in this propaganda."[94] Ferguson, however, was even more explicit about how future responsibilities should be apportioned: "The builder must be progressive and honest. . . . The user should show a degree of tolerance, and sincerity, and willingness to accept advice. . . . The central station management should not only supply current at a reasonable rate, but should also assume that it is to a large extent responsible for the efficient operation of the vehicles."[95] Thus, during the first year or two of the EVAA's existence, those central stations that chose to participate through the EVAA and on their own tried to serve as the coordinating agents for the expansion of the electric vehicle.

The accomplishments of the Electric Vehicle Association of America and the modest expansion of electric vehicle fleets suggested that utilities were better suited to serve as principal agents for the diffusion of electric vehicles than were the vehicle manufacturers. Nonetheless, problems persisted. Several observers recommended the establishment of publicly regulated transportation monopolies (following the same organizational lines as the electric utilities themselves), but this grand scheme always seemed a step too far and was never embraced.[96] A special NELA committee on electric vehicles recommended that "where there is at present no active competition in the selling of electric vehicles, it would be well for the Central Stations to take this matter up and show the possibilities, as has been done in the handling of motors and other special electric appliances."[97] But selling was difficult and time-consuming; and some stations feared that by entering the consumer market directly, they would undermine sales of competing brands and thereby harm their overall charging market. In fact, although central stations were more or less eager to sell the idea of electric vehicle transportation, they remained hesitant about promoting and selling specific vehicle brands. As we have noted, in cities that lacked electric vehicle sales offices, the central station sometimes became a temporary agent for one or more brands. The expectation was that once the business was on firm footing, the central station would withdraw from sales and limit its role to supporting and maintaining all vehicle brands.[98] In reality, however, most central stations, even those whose managers professed enthusiasm for the electric vehicle, were located in cities that did not have active sales bureaus; and despite public assurances to the contrary, few central stations were willing to act as both sales agent, garage operator, and low-cost current provider.

To the extent that manufacturers were making sound vehicles and supportive central stations were providing dependable current at fair prices, responsibility for the growth and maintenance of the electric vehicle fell to the local garage operator. Although Boston Edison opened its Atlantic Avenue garage in 1911, and Edison executive and EVAA president Edward Mansfield argued that "sub-standard" garage service was "a decided menace to the business" in 1911, it was not until 1913 that the garage and the garage man emerged as important loci for reform among EVAA members.[99] As Commonwealth Edison executive R. Macrae described the situation that spring, "without good garage service it is impossible to have good electric vehicle service, and it is now generally recognized that the full development of the electric vehicle industry can be brought about only through the agency of properly equipped public garages."[100] Responding to Macrae and other critics who blamed garages for the slow spread of the electric vehicle, garage owner J. C. Bartlett defended his practices.[101] The garage operator, he claimed, was being held to an unreasonable standard: "We all speak English, of course, but every art, every branch of science or trade has its own language." For Bartlett, the car salesmen promised too much—"fully guaranteed," "anyone can get perfect results," or "a very simple piece of machin-ery"—but when new owners confronted the realities of vehicle op-eration, they blamed the garage.[102] Carroll Haines, another garage owner, framed the dilemma this way: "A garage must charge the bat-tery so that the car will satisfy the agent in respect to mileage, yet the bill for current each month must not be excessive, as well as see that the car owner obtains the maximum life from the battery."[103] Complaints about garage service were not unique to the electric ve-hicle field. Stephen McIntyre's recent dissertation on the history of the automobile service industry shows that concerns about the costs and quality of maintenance have dogged the internal combustion field as well.[104] Nevertheless, because the electric vehicle was more dependent on regular but minor daily garage service than was its gaso-line counterpart, the garage emerged as a target for criticism. By 1915 Harry Salvat, chair of the garage committee and owner of Chicago's flagship Fashion Garage, reported that 95 percent of garage operators "under present conditions . . . prefer the gasoline car business."[105]

The search for a scapegoat for the unsatisfactory growth of the electric vehicle eventually turned to the unskilled laborers who actu-ally operated many of the vehicles. On the one hand, it was thought to be easier to teach a teamster to drive an electric vehicle than a gasoline one. The electric, after all, was supposed to be so simple to operate that even women and children could learn how.[106] Accord-ingly, while wages for an experienced internal combustion vehicle operator ranged as high as twenty-five dollars per week, an electric teamster could be hired for less than half that amount (twelve dollars per week). Whereas the gasoline chauffeur would attend only to the vehicle itself, the driver of the electric vehicle was also expected to help load and unload the vehicle.[107] The electric also promised an additional advantage for fleet owners: because the rated speed of most electric trucks was relatively slow (under twelve to fifteen miles per hour), employers could minimize the wear and tear resulting from a driver who was a "speed maniac." As one *Central Station* correspondent observed, "the only way to clip the wings of the speed maniac is to furnish him with a truck that is geared for low or moderate speed and in which the power is limited, that is to say, furnish him with an electric truck."[108] Thus, on the surface, the electric vehicle offered the possibility of lower operating and depreciation costs.

But cheap labor also meant that workers were not taught about detailed operation of the apparatus. Despite their claimed simplicity, electric vehicles needed limited but regular service and maintenance, and this work was usually assigned to the battery or vehicle expert at the central garage. The operator was so mistrusted that some fleet owners literally forbade the driver to examine a disabled vehicle and insisted instead on a phone call to the garage requesting a rescue. Some in the industry were wary of "giving the impression that the electric vehicle is such a mysterious piece of apparatus." But most would have agreed with Edward Mansfield: when the garage was following a "few simple but important rules," the vehicle was expected to perform "with clock-like accuracy"; under these circumstances, if "you hear of an accident to an electric vehicle, blame the man and not the machine."[109] In response E. H. Freeman of Chicago's Armour Institute of Technology tried to organize a special series of lectures on electric vehicle technology for "practical men engaged during the daytime in electric vehicle or garage work," but the course was cancelled due to lack of enrollment.[110] Several years later, the National Electric Light Association sponsored and offered a similar series of electric vehicle courses in Chicago and San Francisco.[111] But these educational efforts did little to alter perceptions of the vehicle driver as an unreliable component of the electric vehicle system. E. J. Bartlett described typical vehicle drivers as "wild indians," and stories of driver error circulated widely.[112] In a talk before the Motor Truck Club, for example, H. M. Martin described the case of a new electric vehicle that "ran to perfection for two weeks while being operated by the factory demonstrator" but then encountered "constant trouble."

According to Martin, the owner blamed the garage for not charging the vehicle properly when, in fact, the owner's driver had been "running the car with the brake set." It is impossible to evaluate the veracity of such anecdotal accounts, but they do suggest the extent to which the electric vehicle industry struggled to explain the causes of its own troubles. Because the drivers themselves were not represented in the EVAA, they proved a convenient target.

By 1912 the tide of cooperation had turned, and it was the central stations rather than the manufacturers who were feeling abused. Speaking at an EVAA monthly meeting, former association president Blood noted that he was "sorry to say that I have not seen the interest taken in the Association by the vehicle manufacturers that I think should be taken. The central station industry is really bearing the brunt of the association now, and we should get some support and create some interest in the vehicle manufacturers themselves."[113] Later in the year, he called the manufacturers "proverbially a bashful lot . . . [who] never have anything to say, unless it is in the nature of a criticism on the cost of current."[114] A May 1914 editorial in *Electric Vehicles* noted that the expenses of the local sections and national conventions were "all defrayed by the ever ready central station" and took the manufacturers to task for not advertising in the industry trade journal.[115] A cartoon (figure 3.1) captured the mood of the moment; from the perspective of the central station manager, the manufacturers were not doing their fair share and were instead content to let the central stations underwrite the effort to actually sell electric vehicles. John Gilchrist, writing in January 1915 in his capacity as president of the EVAA, tried to reinvigorate the electric vehicle campaign and end bickering in the industry by reminding the various participants of their many areas of agreement. But even he admitted that "cooperation" had "been a much-used word in the electrical industry of late—a word dinned into our ears, indeed, until it has almost lost its meaning."[116] By 1915 meaningful cooperation between central stations and manufacturers was still nowhere in sight.

Clearly, the Electric Vehicle Association of America failed to effectively coordinate the workings of the electric vehicle transport system. There was a need for coordinated intervention to stitch together the fabric of effective electric vehicle service. The EVAA, in partnership with its various affiliated clubs and local chapters, did advance the electric vehicle agenda: it sponsored forums in which ideas and experiences could be shared and did much to highlight the importance of cooperation and coordination for the successful spread of the electric vehicle. But there remained a gulf between knowing what needed to happen and actually establishing a framework within which interests were harmonized. The EVAA and its local sister organizations could not bridge the distance separating the various interests, and talk of cooperation eventually gave way to apportioning blame for the slow growth of the electric vehicle market.

ORGANIZATIONAL COLLAPSE

As the EVAA delegates left the sixth annual convention in Cleveland in October 1915, many must have foreseen the end of the organization as an independent entity. The association had already had various unofficial discussions with NELA about the possibility of merging the EVAA into the larger electrical industry organization. NELA president Edward W. Lloyd, an invited guest at the EVAA convention, had spoken about the possibility of "amalgamation . . . strengthening what should logically be the representative electrical body"; and EVAA treasurer H. M. Edwards had supported the idea, intimating that the continued existence of the independent association was not necessarily assured.[117] At the monthly meeting in November, incoming EVAA president Walter H. Johnson appointed a three-person committee to investigate "ways and means of closer relationship" with NELA, and by March 1916 the consolidation was complete. Although NELA's Lloyd waxed eloquent about the "splendid achievements" of the EVAA, the headline in *Central Station* was less artful: "E.V.A.A. Absorbed by the N.E.L.A."[118] After nearly six years of independent promotion of the electric vehicle, the EVAA had become the Electric Vehicle Section of the National Electric Light Association.

At the first meeting of the Electric Vehicle Section in June 1916, William Kennedy delivered a direct and unequivocal challenge to the electric vehicle industry. Starting with his work as a consulting engineer for the Electric Vehicle Company in the 1890s and the Baker and Studebaker Companies in the 1900s to his continuing involvement with the EVAA and the Society of Automobile Engineers in the 1910s, Kennedy had been party to almost every important electric vehicle initiative. Over two decades, his numerous speeches and

From the Central Station Point of View.

FIGURE 3.1. The changing view of cooperation, 1914. The cartoon illustrates the opinion of many central station operators that vehicle manufacturers were not doing their fair share to market and support the electric vehicle. *Source:* Electric Vehicles 4 (June 1914): 224. Reprinted by permission of the Linda Hall Library, Kansas City, Mo.

papers embodied his unfailing optimism and enthusiasm about the future of the electric vehicle. But in the spring of 1916, he was asking basic questions about the future of the electric vehicle. During the EVAA's almost six years of existence, the total number of electric vehicles in use rose from 8,000 to as high as 35,000. Meanwhile, by 1914 the number of internal combustion vehicles in use had ballooned to more than 1.6 million.[119] Even in the commercial vehicle domain, seemingly the last frontier for the electric vehicle, internal combustion had established a commanding lead. As Kennedy, who had provided so many answers and solved so many problems, addressed the NELA delegates, "fundamental questions" poured forth. Was the central station too preoccupied with "existing activities"? Was the central station in a "free position, legally or logically," to make use of electric vehicles? Was the vehicle business to be "regarded on a par with [other] new business of great magnitude" or classified with "the promotion of sundry lighting and domestic utilities"? In short, Kennedy was asking if the central stations were serious about the electric vehicle proposition. A crisis was at hand, and the electric vehicle business could not "simply float along in a passive state of existence." Even the continued use of the electric vehicle in cities—its supposed separate sphere—was in jeopardy.[120]

The immediate causes of the crisis were twofold: the spectacular growth of the gasoline passenger vehicle market and the impact of World War I on the commercial vehicle. "Even to the casual observer," wrote Kennedy, the gasoline car business had increased in such a "phenomenal manner" that it would eventually "stultify the electric passenger car business." Meanwhile, the "abnormal" growth in demand for gasoline motor trucks for military purposes had allowed gasoline truck makers to unload "old or obsolete materials" and "organize their equipments for quantity production."[121] Truck exports to the European allies had soared during the first two years of the conflict. Some of these vehicles were electric trucks sent, for instance, to Britain to replace gasoline vehicles commandeered for service on the continent, but most were low-priced gasoline models.[122]

No one in the EVAA or the Electric Vehicle Section ever directly admitted that his efforts had failed; there was no concession speech. Indeed, published statements from EVAA representatives were unfailingly enthusiastic and upbeat in their predictions about the future of electric vehicles, even as markets narrowed amid the exponential spread of internal combustion–powered vehicles. Beyond the true howlers (such as EVAA president Smith's claim that the "burden of proof" lay with the gasoline car), EVAA members regularly described the growth of the electric vehicle market as "satisfactory," "gradual," and "steady," or "slow and healthy."[123] Although it is safe to conclude that any vehicle manufacturer would have eagerly traded "gradual" and "steady" for "going gangbusters," EVAA participants nonetheless tried to justify the modest pace of electric expansion. For instance, Stephen G. Thompson of the Public Service Company of New Jersey argued that the average size of electric fleet installations was a sign of confidence in the technology and of the careful tack taken by electric vehicle sales people: "On every hand are found electric vehicle installations whose investment values are of such magnitude that common sense dictates that the selection of type must have been made only after exhaustive investigation. Electric vehicles are not peddled. They are not foisted upon a gullible agent with a demand that he contract for and dispose of a certain quantity with a given time."[124] A few large fleets were taken as proof of success. Other articles claimed that "the electric truck has come to stay" and clung to trivial shreds of statistical data to make their point.[125] Promoters pointed to high annual growth rates, completely neglecting the fact that the total numbers of vehicles in given markets remained insignificant. Arthur Williams, during his term as EVAA president, incorporated data from a survey of commercial vehicles in New Jersey to argue that "nine electric competitors were able to sell more cars than 98 opponents," suggesting that the concentration of production in the hands of a few large manufacturers somehow represented the success of the electric vehicle.[126] One might just as easily have argued that the relative scarcity of electric vehicle producers was a harbinger of imminent collapse. Most curiously, supporters of the electric vehicle clung dogmatically to the idea that electricity, internal combustion, and even the horse each occupied separate spheres of operational efficiency and that each technology would inevitably fail if applied in the sphere of its competitor: "You should cease to think of the electric vehicle as a competitor of the gasoline machine. The true efficiency of the gasoline driven machine commences where that of the electric ends. . . . Whenever either form of vehicle is operated in its wrong sphere, failure is sure to result."[127] Although internal combustion continued to encroach upon the electric sphere throughout the EVAA period, association rhetoric never abandoned its commitment to the principle of separate spheres. Over the six-year life of the EVAA as an independent organization and the five-year existence of the Electric Vehicle Section (1916–20), no more than a handful of published articles even hinted at failure. And as John Lieb's quotation at the beginning of this chapter suggests, pockets of institutional support for electric vehicles survived into the 1920s.

ASSESSING THE IMPACT OF THE EVAA

By 1910 many of the issues that would shape the future of the electric vehicle had already been identified. Provisional battle lines between the vehicle manufacturers and the central stations had been drawn; the benefits of electric vehicle service had been articulated, even as fleet owners were confronting the questions that would continue to plague future electric vehicle operators; and an organizational framework to address these problems and advance the field of electric road transportation was on the horizon. For nearly fifteen years, the central station industry's commitment to the electric vehicle wavered between enthusiasm and indifference. A small group of leading stations, in partnership with vehicle manufacturers and battery makers, tried to pull the rest of the industry onto the electric vehicle bandwagon. But what exactly did the EVAA accomplish? Clearly, domestic interest in the electric vehicle revived in the period immediately before World War I, which coincided with the existence of the EVAA and its affiliated organizations. Nevertheless, establishing that the EVAA was responsible for the electric vehicle revival is problematic.

Since the EVAA focused initially on promoting awareness of the electric vehicle within the central station industry, it is reasonable to evaluate the success of the association by first looking at the extent to which central stations adopted electric vehicles for use in their own service fleets. Tables 3.1 and 3.2 report on two surveys of central station use of electric vehicles—the first in 1909, before the EVAA had organized its national campaign; the second in 1913, far enough into the life of the organization to allow a rough estimate of the effects of its efforts. Both the absolute size and relative ranking of the electric vehicle fleets in central station service in the given cities are reported in table 3.1, while table 3.2 lists all the cities reporting in the 1913 survey ranked by central station fleet size along with the number of passenger electric vehicles in use in the respective service areas.

Table 3.1 Rank and absolute size of electric vehicle fleets in central station service, 1909 and 1913

CITY	1913		1909	
	Rank	Number	Rank	Number
New York, N.Y.	1	125	1	58
Chicago, Ill.	2	103	3	17
Boston, Ma.	3	86	6	4
Detroit, Mich.	4	86	8	2
Rochester, N.Y.	5	62	5	5
St. Louis, Mo.	6	41	4	10
Topeka, Ks.	7	40	–	–
Philadelphia, Pa.	8	38	2	18
Bridgeport, Conn.	9	35	–	–
Baltimore, Md.	10	30	–	–
San Francisco, Calif.	11	29	–	–
Denver, Colo.	12	24	–	–
Los Angeles, Calif.	13	21	–	–
Lawrence, Ma.	14	17	–	–
Hartford, Conn.	15	14	–	–
New Orleans, La.	16	12	–	–
Jersey City, N.J.	17	9	–	–
Kansas City, Mo.	18	9	–	–
Cambridge, Ma.	19	8	–	–
Schenectady, N.Y.	20	8	9	2

Sources: "The Relations between Central Stations and the Electric Automobile," *Electrical World* 54 (August 5, 1909): 314–20; *Proceedings of the Fourth Annual Convention of the Electric Vehicle Association of America* (Chicago, October 27–28, 1913), 14–26.

Tables 3.1 and 3.2 show that most of the top ten companies in 1913 had been leading users of electric vehicles in 1909. Central stations in New York, Chicago, Boston, Detroit, Rochester, St. Louis, and Philadelphia operated electric vehicles before the creation of the EVAA; indeed, these companies had organized and supported the association from its inception as the Electric Vehicle and Central Station Association. From 1909 to 1913, average fleet size at the leading companies grew, and there were some shifts in the rankings but little overall change at the top; these companies did not need encouragement from the EVAA to adopt electric vehicles.

But the data also reveal the emergence of a second tier of central stations that had not reported owning and using electric vehicles in 1909. These newcomers can be subdivided into several smaller groups. Between 1909 and 1913, the rapidly expanding metropolises of the far west—Denver and Los Angeles especially but also San Francisco to a lesser extent—added electric vehicles in central station service and demonstrated relatively high use of electric passenger vehicles. For these cities, regardless of the absolute cost-effectiveness of their own electric vehicles, promoting the electric vehicle may well have resulted in profitable sales of current to owners of passenger vehicles. Local sections of the EVAA were established in each of the three western cities, but causality is unproven. A second group—small midwestern cities such as Akron, Lincoln, Peoria, South Bend, Topeka, and Wichita—had also responded to the message of the EVAA. Although absolute fleet size in these areas did not approach those in the major cities and no local EVAA sections were established, the population-adjusted rankings show that central stations used electric vehicles and that passenger electric vehicles were also relatively successful. The tables also indicate that smaller eastern cities such as Baltimore and Bridgeport and independent suburbs such as Cambridge, Massachusetts; Jersey City, New Jersey; and Pawtucket, Rhode Island, had imitated the larger central stations in New York, Boston, and Philadelphia by purchasing electric service vehicles. These smaller cities and suburbs, however, did not report increased use of passenger electric vehicles. Finally, the national picture (figure 3.2) confirms that average reported fleet size increased dramatically from 1909 to 1913. Mean fleet size doubled from just over five to twelve, and median fleet size increased from one to four vehicles. But response to the electric vehicle campaign was uneven. Reports were received from fewer than 75 central stations out of more than 5,000 operating in the country; and even near Boston, where the Edison Company had committed considerable resources to increasing the visibility and appeal of electric vehicles, manufacturer William Anderson concluded, as late as 1913, that many stations "are still living in the Dark Ages, quite content to operate their business to-day along the same lines followed by their great grandfathers . . . [and] that there are localities within fifty miles of where I stand [Boston] that are charging 25¢ per kilowatt hour for electric charging current."[128] Taken together, the available data do show that American central stations increased their use of electric vehicles from 1909 to 1913; but without detailed accounts of firm-level decision making, it is not possible to establish the EVAA's specific role in causing or accelerating this process.

The problem of causality is even more pronounced when trying to connect the EVAA to the revival of interest in electric passenger vehicles more generally. In a 1913 speech, EVAA president Arthur Williams, commenting on the modest growth in the use of electric vehicles, claimed that the association was "justified in attributing a

Table 3.2 Electric vehicle fleets in american cities, 1913, listed in order of central station fleet size

	CENTRAL STATION FLEET				PASSENGER VEHICLE FLEET			
			Adjusted by Population				Adjusted by Population	
	Rank	Absolute Number	Rank	Absolute Number (per 1,000)	Rank	Absolute Number	Rank	Absolute Number (per 1,000)
New York, N.Y.	1	125	41	0.026	14	225	62	0.047
Chicago, Ill.	2	103	19	0.062	1	3075	18	1.854
Detroit, Mich.	3	86	8	0.125	4	810	12	1.179
Boston, Ma.	3	86	31	0.039	15	214	40	0.098
Rochester, N.Y.	5	62	6	0.284	66		66	0.000
St. Louis, Mo.	6	41	20	0.061	6	550	27	0.820
Topeka, Ks.	7	40	7	0.178	37	39	25	0.174
Philadelphia, Pa.	8	38	12	0.082	11	254	51	0.545
Bridgeport, Conn.	9	35	2	0.481	41	25	47	0.343
Baltimore, Md.	10	30	3	0.467	48	15	63	0.234
San Francisco, Calif.	11	29	11	0.088	19	168	37	0.507
Denver, Colo.	12	24	1	0.549	3	822	3	18.811
Los Angeles, Calif.	13	21	4	0.369	2	2000	1	35.163
Lawrence, Ma.	14	17	25	0.056	57	5	59	0.017
Hartford, Conn.	15	14	5	0.320	12	250	6	5.723
New Orleans, La.	16	12	9	0.118	40	30	56	0.294
Jersey City, N.J.	17	9	51	0.022	16	204	28	0.489
Kansas City, Mo.	17	9	61	0.016	8	530	10	0.949
Schenectady, N.Y.	19	8	33	0.037	52	13	50	0.061
Cambridge. Ma.	19	8	44	0.025	44	20	49	0.063
Washington, D.C.	21	7	13	0.081	5	756	8	8.802
Salem, Ma.	22	6	21	0.061	48	15	39	0.152
Indianapolis, Ind.	22	6	46	0.024	6	550	7	2.214
Binghampton, N.Y.	22	6	49	0.022	31	75	16	0.280
Minneapolis, Minn.	22	6	59	0.018	9	500	13	1.475
Memphis, Tenn.	26	5	10	0.103	64	0	64	0.000
Providence, R.I.	26	5	27	0.048	41	25	53	0.238
Dayton, Ohio	26	5	52	0.021	21	150	21	0.642
Saginaw, Mich.	29	4	14	0.079	26	100	11	1.980
Pawtucket, R.I.	29	4	16	0.077	57	5	54	0.097
South Bend, Ind.	29	4	17	0.075	33	60	23	1.118
Harrisburg, Pa.	29	4	34	0.034	44	20	41	0.172
Omaha, Nebr.	29	4	37	0.032	17	195	15	1.571

Table 3.2 *Continued*

	CENTRAL STATION FLEET		Adjusted by Population		PASSENGER VEHICLE FLEET		Adjusted by Population	
	Rank	Absolute Number	Rank	Absolute Number (per 1,000)	Rank	Absolute Number	Rank	Absolute Number (per 1,000)
Brockton, Ma.	29	4	39	0.031	48	15	45	0.114
Portland, Ore.	29	4	57	0.019	34	55	44	0.265
Pittsburgh, Pa.	29	4	66	0.007	12	250	35	0.468
Dubuque, Iowa	37	3	15	0.078	61	3	57	0.078
Joplin, Mo.	37	2	18	0.062	48	15	36	0.468
Allentown, Pa.	37	3	22	0.058	60	4	58	0.077
Auburn, N.Y.	37	2	23	0.058	54	10	42	0.288
Wichita, Ks.	37	3	24	0.057	36	42	26	0.801
Elmira, N.Y.	37	2	26	0.054	54	10	43	0.269
Lincoln, Nebr.	37	2	28	0.045	29	95	9	2.16
Peoria, Ill.	37	3	29	0.045	10	260	2	3.88
Akron, Ohio	37	3	30	0.043	18	185	5	2.67
Mobile, Ala.	37	2	32	0.039	37	39	29	0.757
Lynn, Ma.	37	3	35	0.034	44	20	48	0.224
Dallas, Tex.	37	3	36	0.033	24	115	22	1.249
Savannah, Ga.	37	2	38	0.031	26	100	17	1.537
Duluth, Minn.	37	2	43	0.025	39	38	34	0.484
New Haven, Conn.	37	3	47	0.022	53	12	55	0.090
Atlanta, Ga.	37	3	55	0.019	26	100	32	0.646
Nashville, Tenn.	37	2	58	0.018	32	74	30	0.671
Seattle, Wash.	37	3	63	0.013	20	155	31	0.653
Woonsocket, R.I.	55	1	40	0.026	62	2	60	0.052
Montgomery, Ala.	55	1	42	0.026	54	10	46	0.262
Superior, Wis.	55	1	45	0.025	62	2	61	0.050
Chattanooga, Tenn.	55	1	48	0.022	57	5	52	0.112
Little Rock, Ark.	55	1	50	0.022	43	24	33	0.522
Lancaster, Pa.	55	1	53	0.021	47	18	38	0.381
Canton, Ohio	55	1	54	0.020	35	50	24	0.996
Springfield, Ill.	55	1	56	0.019	21	150	4	2.903
Terre Haute, Ind.	55	1	60	0.017	30	80	20	1.376
Evansville, Ind.	55	1	62	0.014	25	110	14	1.579
San Antonio, Tex.	55	1	64	0.010	23	135	19	1.397
Fall River, Ma.	55	1	65	0.008	65	0	65	0.000

Sources: Proceedings of the Fourth Annual Convention of the Electric Vehicle Association of America (Chicago, October 27–28, 1913), 14–26.

large share of this prosperity to our organization's efforts. . . . Cooperation, then, is the force which has been bringing the electric motor car into its own, and the plan for this cooperation originated with the EVAA."[129] According to existing data, such as those reported through the 1913 EVAA survey of central stations, size of central station electric vehicle fleet was weakly correlated with the number of electric passenger vehicles reported in a given area. It is also possible, however, that the causal relationship was reversed and that, in some instances, the broad revival of interest in the electric vehicle spurred central stations into action. Regardless, for purposes of evaluating the Electric Vehicle Association of America, the relationship among EVAA organizing, central station electric vehicle fleet growth, and the use of electric passenger vehicles suggests that organizational support for the electric vehicle contributed to the success of the electric vehicle in the years before World War I.

Moreover, the EVAA, through its long-running debates about cooperation and selling, helped raise the visibility of transportation service within businesses that had long taken their stables for granted. The very idea of a transportation engineer—a person whose job was to look closely at the real costs of transportation service—resulted in part from the EVAA's work. Because electric truck makers were ahead of their gasoline counterparts in developing this perspective, the advent of the commercial electric vehicle represented the first time that important capital allocation decisions were being made in the stables. Like the rise of information systems in modern corporations, the rising cost and increasing importance of motorized transport altered relations within firms and brought greater responsibility and accountability to the staff of the transportation division.

In the end, despite the fact that the electric vehicle was unable to hold its own against competition from internal combustion, the EVAA and the electric vehicle revival produced few losers. The battery makers grew and profited with the electrification of the gasoline car. Demand for electric current continued to expand, and the position of the central stations vis-à-vis the public and the state produced a stable and profitable trajectory of growth. Suppliers such as General Electric hardly noticed the absence of the electric vehicle market. For instance, mercury-arc rectifiers, initially applied in small but efficient home charging units for electric vehicles, were simply made larger and larger to suit the operating scale of the rest of the electrical industry.[130] Only the electric vehicle manufacturers de-

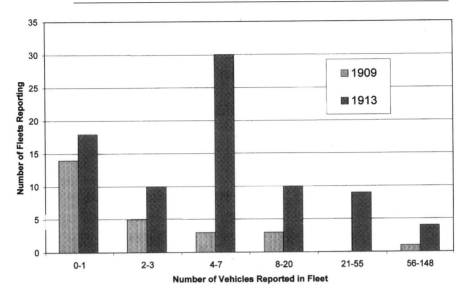

FIGURE 3.2. Electric vehicles in central station service, 1909 and 1913. *Sources: "The Relations between Central Stations and the Electric Automobile,"* Electrical World *54 (August 5, 1909): 314–20;* Proceedings of the Fourth Annual Convention of the Electric Vehicle Association of America *(Chicago, October 27–28, 1913), 14–26.*

pended on the success of either the commercial or the passenger electric vehicle. Along with 99 percent of the companies making gasoline vehicles, many electric manufacturers failed to survive the seismic changes that shaped the modern automobile industry. But even here, there were exceptions. Demand for industrial trucks continued to expand; and the Baker Company, established by Walter Baker in 1899, when he drove his first electric vehicle through the streets of Cleveland, merged with local competitor Rauch and Lang in 1915 and gradually shifted its focus to the industrial equipment sector.[131] The company continued to produce electric industrial trucks until the 1970s, when its European owners shut the doors of the Cleveland plant.

Throughout the existence of the EVAA and its companion organizations, enthusiasm among participants rarely flagged. How could presumably intelligent, well-informed executives have clung so tenaciously to their hopes for the electric vehicle? Although several related explanations are possible, let us dispense with one. After scouring their meeting records, I find no evidence of deceit. EVAA representatives may have massaged data to suit their pedagogical aims, but there is no sign that the leadership was any less deluded about the prospects for the electric vehicle than indicated by their public posturing. A more plausible explanation draws on the institutional culture of the central station industry during the second decade of the century. Speaking to the delegates at the NELA convention in 1909, T. Commerford Martin, editor of *Electrical World*, reminded his audience that in 1886 he had spoken before a National Electric Light Association consisting of only twenty electric lighting companies and that the following year he had read a paper on electric traction before only thirteen members of the American Street Railway Association. "Today," he concluded, "there are 5000 central stations and 1300 electric railway systems in this country, and it is apparent that the future of electric vehicle transportation is equally bright."[132] The central stations were unaccustomed to failure. As historian David Sicilia observed, for Boston Edison the electric vehicle campaign was "probably the least successful large-scale marketing effort of the period."[133] Because all of their other promotions had produced favorable results, EVAA leaders may have been too patient, expecting that eventually their efforts would meet with success. The spectacular growth of the internal combustion vehicle market—described in the pages of *Central Station* as "one of the wonders of the age"—must have further emboldened the EVAA. As Frank Stone, the association's first president commented in 1909, "we see hundreds of horse-drawn vehicles which we are pretty certain *will not be seen* a few years hence. *On that point, I think all of us agree.*"[134] To the founders and organizers of the association, it seemed perfectly plausible to imagine a shared, hybrid transport system with separate spheres of economic action for each of the various prime movers. Even if the separate sphere for the electric were ultimately quite small, given the size of the total vehicle market, it still looked promising. Only if the electric vehicle sphere were reduced effectively to zero would the association have been convinced to scale back its efforts. By most standards the electric vehicle industry experienced solid and significant growth during the EVAA years. It was only in contrast to the spectacular expansion of the internal combustion sector that the growth of the electric vehicle looked so anemic. Finally, the central station executives came from the generation of the electric prophets. Their faith in the future of electricity was unshakable. As one EVAA member wrote, "if you dam up electricity or any other primal force in one direction, it will overflow in some other. The longer you keep it pent up the more pressure you generate. Great movements

gather momentum slowly and their significance is often underestimated by the world at large."[135] For EVAA members, it was simply a
matter of time before their primal force would overcome the barriers
in its way.

Another issue concerns the timing of the central station initiatives. Why did it take ten years for the electrical industry to awaken
to the vehicle opportunity? Speaking in 1914, long-time electric vehicle advocate William Kennedy struggled to understand why it had
taken the central station industry so long to recognize the potential
of the electric vehicle market. Given the "progressiveness of this industry in comparison with others," he said, it was nothing short of
"remarkable" that the electric vehicle had remained "latent for such
a lengthy introductory period."[136] In retrospect, several explanations
are possible. Richard Schallenberg attributes the delay to the decline
of interest in passenger electric vehicles. Only with the commercial
revival did the central stations rediscover the merits of the electric
vehicle load. Of course, this interpretation misses the point that the
electric vehicle revival occurred in part because a few central stations
never stopped using electric vehicles. Perhaps the successful introduction of the Edison iron-nickel battery explains part of the change of
heart among the central station executives, who had been trained
with Edison's distrust of the lead-acid battery ringing in their ears.
Looking at the development of various industrial sectors, we can see
that the electrical industry was still in the midst of major structural
consolidations and technological changes that may have blinded
management to the electric vehicle market. In 1900 central station
managers were focused on increasing scale, expanding production,
and building larger and more efficient electrical systems. The first
turbogenerators led to larger turbogenerators, not to the small, distributed systems used in automobiles. To many in the industry, the
car was still just a toy—a weekend distraction—not a serious business proposition. It was only as the central station business environment stabilized and managers began to see the automobile as more
than another fad that the industry began to look seriously at the electric vehicle market.

Finally, might the competition among steam, gasoline, and electric vehicles have turned out differently if the central station industry had created the EVCSA in 1898 instead of 1908? Possibly. Writing
about the EVAA, historian David Sicilia has concluded that "marketing could not guarantee the economic success of a weak technology;

at best it could only induce an Indian summer." But had the association existed in the late 1890s, when all three technologies were, in certain respects, equally "weak," concerted intervention by a powerful, unified industry might have tipped the scales toward a more robust separate sphere for the electric vehicle. Instead, a decade-long head start was too much for the central station industry to overcome.

4

THE ELECTRIC TRUCK
AND THE RISE AND FALL
OF APPROPRIATE SPHERES

Here is a truck. It is potentially efficient. If you know how to
run it, take it out and save money.

Editorial, *Commercial Vehicle*[1]

How did commercial conditions for electric trucks differ from those
that shaped the market for the private passenger automobile? During the transition from horse wagon to motor truck, functional specialization—the idea that each motive power was uniquely suited to
perform a particular set of tasks—continued to shape the market for
commercial vehicles long after the passenger car market had settled
on internal combustion. Whereas the pleasure vehicle quickly created new markets, the commercial vehicle was only gradually incorporated into existing, horse-paced business practices. Early users of
commercial vehicles operated in a short-haul service environment.
Distinct zones of economic operation were consistent with the
operating parameters established by the horse-paced world but were
difficult to define and almost impossible to maintain over time. Although the Hartford Electric Light Company's 1912 introduction of
battery exchange service offered a solution to many of the problems
facing electric trucks, battery service failed to resolve the internal conflicts in major urban markets among different vehicle producers and
battery makers. Paradoxically, therefore, battery service signaled the
electric vehicle industry's acceptance that the electric truck was incapable of maintaining its separate sphere.

The outbreak of World War I ended the broad resurgence of interest in the electric commercial vehicle. Impelled by skyrocketing
demand for wartime exports, the capabilities of the gasoline truck

expanded, along with the production capacity of the American truck industry. The horse and the electric vehicle were left to struggle for a shrinking share of the domestic truck market. By the end of the war, the military applications of the gasoline truck had firmly established both demand for internal combustion and support for peacetime road building. After the war, the appropriate sphere of the electric truck shrank to a handful of niche applications, whereas the opportunity for internal combustion trucks continued to grow.

THE EARLY MARKET FOR THE COMMERCIAL MOTOR VEHICLE

Today, the idea of trucking conjures images of massive trailer trucks criss-crossing the country over the interstate highway system in competition with the nation's railroads.[2] But long-haul freight motorization was the last reaction, not the first, to the impact of the railroad. During the five decades between the end of the Civil War and the outbreak of World War I, railroad expansion had allowed more and more goods to move increasingly cheaply from railhead to railhead. The transport challenge of the rail era was local collection and distribution. Local companies used thousands of horses to carry goods the last mile or two from the railhead to the distributor. In practice, this horse-based, short-haul service cost as much or more per pound than the entire long-haul journey. As F.M.L. Thompson has described in the case of turn-of-the-century London, the railroads, along with the symbiotic express companies that flourished at the edges of the rail systems, were the largest owners of horses.[3] Accordingly, short-haul trucking was the first arena in which the commercial motor vehicle had to prove its worth.

For most early twentieth-century American businesses the capabilities of the horse-drawn wagon defined efficient transport. Regional wholesale producers—bakers, brewers, ice houses, and coal distributors—used large (five-ton) wagons to deliver their products to local merchants. Similarly, local merchants and the growing, market-hungry department stores used small (1,000–pound) one-horse wagons to deliver merchandise to individual customers. Efficient transport service was also based on prevailing modes of commerce: people bought and sold locally. Delivery was the norm rather than the exception; before the introduction of parcel post in 1913, merchants provided delivery service free of charge. Some businesses subcontracted deliveries with local transport providers, but leading merchants frowned on the general idea of cooperative delivery. As C. R. Langenbacher, delivery supervisor of Lord & Taylor department store in New York, said, "How would it look for a delivery boy to enter an apartment or a house with a lady's hat in one hand and a ham gracefully poised in the other?"[4] Delivery offered an opportunity for the merchant to maintain direct contact with the customer, and many were unwilling to entrust its provision to contractors.

Although the idea of using motor vehicles to serve commercial ends was not new, the commercial vehicle market developed after the market for private passenger vehicles. As of 1903, a handful of companies was already producing custom-designed commercial vehicles, but most of these designs were derived from those of pleasure vehicle chassis.[5] But as H. F. Donaldson, editor of *Commercial Vehicle*, noted in his inaugural editorial, by 1906 important differences between the pleasure and commercial vehicle markets had begun to emerge. Whereas the private car could "make a poor showing on the private ledger of its owner and still be considered an unqualified success," the commercial vehicle must "make good in dollars and cents. . . . It is a machine that makes money for its owner."[6] Private vehicles were operated only during fair weather and were often garaged for months at a time, but commercial vehicles needed to work six days per week, fifty-two weeks per year. If a vehicle could not run through all weather and road conditions, its value as a replacement for horses was limited. Commercial vehicles also had to contend with the "driver problem": whereas pleasure vehicles were usually driven by the owner or under the owner's direct supervision, commercial vehicles were often entrusted to unskilled or untrained employees. The *Boston Journal* summed up the difference: "the touring car is a money spender," but the commercial car should be "a money earner."[7]

During the first years of the new century, motor trucks were employed where they were most visible and where their service reflected most favorably on the owner. Beyond advertising, however, it was not immediately clear how best to take advantage of the new capabilities that the truck offered. A number of specialized vehicles were developed for narrow market niches. In New York City, for instance, piano delivery was much easier by electric truck because, in addition to moving the instrument from warehouse to customer, the battery could power a winch to lift the piano into the apartment or house. Similarly, central stations fitted electric trucks with a lighting tower (a flying bridge from which the driver could both operate the vehicle and adjust the street lamp lights), thereby obviating the need for a second teamster to steady the wagon.[8] Still, these specialized vehicles served limited markets.

By 1906 users of commercial motor vehicles could be divided into two categories: express service and bulk delivery. The express companies that transported small and medium-sized packages to and from railheads and dockyards owned some of the largest stables in American cities and experimented with many different types of motor vehicles. Purchasing trucks in lieu of fresh horses, express companies "stabled" them along with the animals and used them initially on the service routes of the horses they replaced. Leading local merchants—B. Altman, R. H. Macy's, Gimbel's, Tiffany, Abraham and Straus, and Gorham in New York and Houghton & Dutton in Boston—were among the first companies to purchase commercial vehicles.[9] Generally, these vehicles replaced one- or two-horse wagons and carried packages totaling no more than 1,000 pounds. Both electric and gasoline vehicles were faster than a horse; both were more modern and progressive than the horse wagon; and both could be parked on the street, thereby liberating stable space for other corporate uses.[10] Nevertheless, both electric and gasoline vehicles were more expensive to purchase and operate than were horses and wagons; and many early users soon recognized that motor vehicles, whatever their motive power, would have to do more than simply replace horse wagons on a one-for-one basis.

Motor vehicles also made inroads into the market for the delivery of bulk supplies. Coal, ice, and beer distributors were typical early customers. Breweries used five-ton motor trucks to deliver fresh beer to local taverns.[11] The distances covered were usually small—often fewer than five miles for a given route—and speed was much less important than was dependability of service. On an average day, horse-drawn vehicles were perfectly capable of doing the job. It was the atypical day, however, that was most important. When the weather was hot, ice melted faster, and people consumed more beer; yet the heat also exhausted the horses and rendered them incapable of providing their normal service, let alone carrying more weight or covering their routes a second or third time. Similarly, during a prolonged cold snap or following a major winter storm, demand for heating coal rose at the same time that horses froze in their stables and failed to pull through unplowed city streets. A Chicago coal delivery company, for instance, advertised the fact that its five-ton gasoline truck had managed fourteen 5.3–mile trips on a single January day in 1912 when the temperature was twelve degrees below zero. Per-mile costs were irrelevant because no horse team would have survived such an ordeal.[12] As a survey article in *Horseless Age* observed, "ear-lier means of cartage and transportation offer[ed] no parallel" for such applications.[13] These businesses valued the motor truck not for its speed or its strict economy of operation but for its ability to perform under a wide range of adverse conditions.

FROM HORSE PACE TO SEPARATE SPHERES

Who favored which type of commercial vehicle? For distributors of bulk commodities, steam and electric vehicles were the early favor-ites. At the Liverpool commercial vehicle trials in England in 1901, steam dominated the field.[14] Although steam remained popular with British commercial customers, vehicles based on similar designs found only limited success in the American market. As in the pleasure ve-hicle market, steam rapidly disappeared from the urban, short-haul vehicle market.[15] Compared to the situation for private passenger vehicles, commercial vehicle standards were slower to emerge. But given the astounding growth of the gasoline private car, expectations for the internal combustion truck were high. Commenting on the results of the first American commercial vehicle contest, held in New York City in May 1903, *Horseless Age* contributing editor Albert Clough concluded: "Experience will furnish the decisive evidence which shall assign each motive power to its appropriate sphere. Light delivery work in congested areas will perhaps become the province of the electric vehicle. In sparsely settled territory this duty may de-volve upon the gasoline engine, while steam may do the bulk of the heavy trucking, pending the application of internal combustion en-gines to this service."[16] Clough, who was affiliated with a leading journal and was not himself staked to the success of any one of the motive powers, thus enunciated the idea of appropriate or separate spheres.

Frederick Winslow Taylor's famous time-and-motion studies ush-ered in a new approach to the organization of work that stressed rep-etition, specialization, and efficiency. As more and more of the economy "Taylorized," the horse-based transport system cried out for reform.[17] Motorization offered a wedge into the tradition-bound realms of the stable and the shipping room. Although some merchants instituted organizational changes without switching motive power (one claimed to have "saved a cent" per package simply by reorganizing the loading department), most businesses reorganized and motorized at the same time. Motor trucks, with their high initial and operating costs and their promise of increased performance, were used to justify changes in shop practices.[18] New accounting and tracking systems allowed owners to measure performance and combat "shipping room constipation."[19] A 1914 editorial decried the two cardinal sins of motor trucks in short-haul work: "the horse-pace blunder" and the "moving-factor error," referring to the high percentage of total time spent standing rather than moving. The unidentified company that was the subject of the editorial had increased the overall efficiency of its delivery service by canceling transfer service, closing remote substations, and using motor trucks for direct delivery to distant customers.[20] The first, efficiency-driven wave of motorization is well illustrated by the two maps in figure 4.1. A hypothetical horse-based delivery system is shown in the first panel. The second shows the delivery pattern after three motor trucks have been added to supply new substations. The trucks allowed more frequent deliveries in the immediate neighborhood of the store and extended the overall area of coverage.[21] But initially trucks were not expected to do away with the need for horses, merely to improve the efficiency of the overall transportation department. As an editorial in 1907 in the *Chicago Post* observed, "it is too soon to lament the horse. We have not yet come to the day when we must decide whether we are to pet him, as the dog, or eat him as the amiable cow. Our sentiment for the noble beast will remain, and with the heaviest work undertaken by insensitive strength [motor vehicles] his lot will probably be improved."[22] Motorization was first seen as a component of efficient organization, not as a goal in itself.

Combining traditional transport practices with expensive and unproven technology, in effect, created separate spheres of action for each technology. The first commercial vehicle buyers bought trucks to replace some of their horses or to augment their horse-drawn fleet. As late as 1914, few companies had dispensed with all of their horses.[23] Given that owners were still paying for stabling, the horses were presumably being used; and because even the most conservative owner would have realized that horse wagons and motor trucks should not be used interchangeably, every owner of a mixed fleet

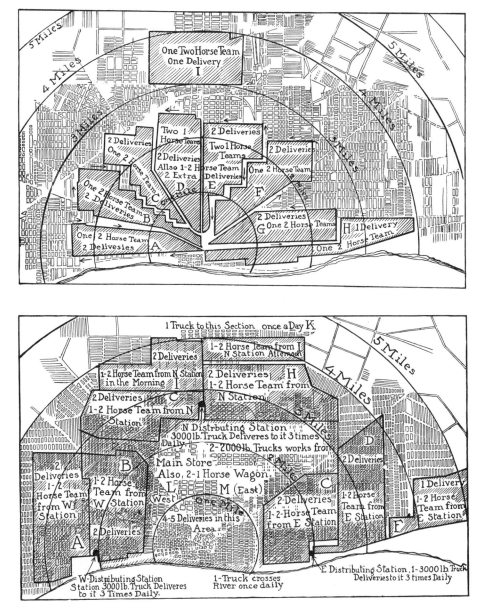

FIGURE 4.1. Hypothetical distribution system before and after efficiency-based motorization, 1912. *Source: "What Becomes of the Wagons and the Horses?"* Commercial Vehicle 7 *(February 1912): 38–39.*

would have been forced to apportion specific duties to different types of vehicles. The gradual process of motorization created separate spheres; the remaining question was simply which vehicle was best suited to what type of work.

Separate spheres for horse wagons and motor vehicles were rooted in "horse pace"—the way turn-of-the-century companies organized their transport service. Having used horses for decades, businesses had adopted a host of practices that were co-adapted to the capabilities of the horse. It is impossible to catalog all of these practices, but some examples are instructive. Like all beasts of burden, horses were subject to fatigue. Over the course of a day's work, a horse team might cover eighteen to twenty miles at an average pace of three to five miles per hour; however, an experienced teamster would provide frequent stops for water and rest to avoid pushing a team to exhaustion. Periods of rest would allow the team to perform at full power the following day. Either intentionally or accidentally, business practices had evolved to create such rest stops. Delivery service, for instance, usually offered customers the option of paying for goods at the time of delivery (c.o.d.). Patrons were also allowed, even expected, to open and inspect the contents of all packages. Clothing and other personal items were tried on while the team, as well as the driver or delivery boy, waited. Similarly, in heavy brewery service, the team operator was responsible not only for unloading the new barrels and removing the empties but also for tapping the fresh keg. Tradition then dictated that the driver sample the product with the tavern owner.[24] All the while, the team of horses was recuperating from hauling the heavy load. Business methods had evolved to accommodate the physical attributes of the horse.

Unlike horses, motor vehicles did not generally need time to recuperate between loads; and given the initial purchase price of a truck and the fixed costs associated with its maintenance, waiting time was lost time. With the spread of motorized commercial vehicles, truck drivers did not stop observing time-honored practices such as drinking a mug of beer with the tavern owner or waiting for customers to model a new item, but the opportunity cost—the foregone service—associated with such behavior increased dramatically. Joseph Husson's study of beer delivery in New York City, for instance, reported that even after motorization, 15 percent of total work time was spent "partaking of liquid refreshment."[25] The examples of beer delivery and package service underscore the extent to which the potential efficiency of motor truck service was compromised by continued adherence to horse-paced transport practices. As noted in an anonymous column in *Electric Vehicles* titled "Choosing the Type," "[a] little rest now and then for a team of horses is good, but with a motor truck, it is not only unnecessary, but unprofitable."[26]

Horse-pace, however, also helps explain the peculiar persistence of the appropriate sphere of the electric truck. In its various operating characteristics, the electric vehicle was more similar to the horse than was the gasoline truck. The electric truck traveled faster than the horse wagon but not so fast to encourage joy riding and overspeeding. As Clinton Woods observed in 1897, an electric motor could be "momentarily worked up to two and three times its normal rated power" in much the same way that a horse could be coaxed or tortured into pulling a little harder.[27] Confronting the same hill or curb, an internal combustion engine would simply stall. And a battery, like a horse, could do with a short rest now and then. Most of all, if stores or distributors wanted to maintain the approximate outline of their distribution system and still increase overall delivery range to keep up with an ever-expanding customer base, the effective operating range of the electric truck was greater than that of the horse wagon. In contrast, the efficient use of gasoline trucks required that whole service and delivery organizations be reorganized around the new technology to allow longer delivery routes. Defending the need to completely rebuild transportation departments, a 1907 editorial pleaded that "transportation by machine must be on a machine basis":

> To buy a truck and operate it on a horse schedule would be as sensible an undertaking as to operate a trolley road on the schedule of the horse car it superseded. . . . Those in the trade are not merely bettering an existing form of transportation service, we are all engaged in the work of applying the resources of modern mechanics in complete substitution for the antiquated and inadequate system of animal traction, as radical a change in its way as that which advanced from the tallow candle to the incandescent lamp. . . . To make *good* machines is not sufficient; to insure that they will be put to *good use* is quite as important.[28]

Easier said than done. It would take many years to accomplish this "complete substitution," and in the interim the electric vehicle continued to have an important role. "To adapt the [gasoline] truck to the shipping room is like making the man over to fit the coat," observed a 1913 editorial in *Commercial Vehicle*.[29] But for the risk-averse manager of the typical delivery department, it was easier to change only one variable at a time and let electric trucks do a little more than horse wagons. The electric truck was a new coat that already fit. Rather than let gasoline trucks dictate large-scale organizational changes, some managers adopted an incremental approach. Full-scale motorization with gasoline trucks and the accompanying organizational changes would come in due time, but many companies used the electric vehicle as their first step toward motorization.

THE LIABILITIES OF INTERNAL COMBUSTION

Although the gasoline car had proven its value in the private pleasure car market and internal combustion had the potential to provide universal commercial service, owners of early gasoline trucks encountered a host of problems. Unpredictable repairs and high maintenance costs resulted in unreliable service. As A. N. Bingham observed in an article on the operating costs of gasoline motor trucks, "excessive repairs ruin an owner's disposition and a manufacturer's reputation. . . . Reliability is the manufacturer's greatest asset."[30] Four years later, anxieties about this "asset" remained: the commercial motor truck was described as "a game of bluff in which the seller stakes his reputation against the dubious ability of the user to take proper care of the vehicle."[31] For many commercial customers, the fear of breakdowns was worse than living with old Dobbin.[32] And even in service, the gasoline trucks were problematic. Because internal combustion motors were difficult to start and expensive to leave running, gas trucks were not well suited for the stop-and-go delivery service that made up the majority of urban commercial work. Electric starters, first introduced on passenger cars in 1911, failed to address this situation because most manufacturers thought a starter for a truck motor would have to be too heavy and bulky to be worthwhile. With fifty or more starts per day, the truck would need to be equipped with a large storage battery.[33] Moreover, given concerns about reliability, truck makers were hesitant to add yet another subcomponent that might malfunction.[34] The net result was that automatic starters spread relatively slowly to gasoline commercial vehicles. By model year 1917, only slightly more than half the models in production included starters (56.33 percent), and fully one-sixth had no electrical systems at all.[35]

A second family of problems derived from the "driver question," an important factor guiding merchants' technological choices. Hired hands had no personal interest in protecting the capital assets of vehicle owners. Although in principle driver agency was a problem for both electric and gasoline trucks, in practice the higher speeds attainable with internal combustion meant that drivers of gasoline trucks had more opportunity to damage a truck and its cargo than did their electric counterparts. As one observer admitted, "when an operator finds himself 25 miles from home with his car unloaded and the supper hour within sight, it is too much to hope that he will return at anything but the maximum speed of which the car is capable." Reviewing the "problem of drivers" in 1905, *Horseless Age* wistfully concluded: "If operators were obtainable who would take the same interest in the cars that they are hired to drive that they would if they were their own property, the problem of the commercial motor would be greatly simplified."[36] Yet interest alone might not have been sufficient. As E. J. Kilbourne observed in 1914, the typical driver was an ex-teamster who had never operated a motor vehicle. The temptation to experiment—"to determine for his own curiosity exactly how much ability exists in the vehicle, both in speed and propelling power"—ensured that the vehicle was "subject to considerable abuse" during its initial trials. Even when this "personal failing" was not present, lack of experience often led drivers to operate vehicles in ways that were "detrimental yet unknown."[37]

As roads and vehicle technology improved and delivery routes lengthened, speeding (or "overspeeding") became endemic. In 1910 even twenty miles per hour on solid tires over a rough surface resulted in frequent mechanical breakdowns and increased tire wear. An article titled "Trend of the Times" counseled that doubling average speed from eight to sixteen miles per hour trebled tire wear.[38] A prevailing rule of thumb held that it was false economy to travel faster than six to eight miles per hour.[39] Enumerating the liabilities of the gasoline truck at an EVAA meeting, E. J. Bartlett claimed that "speed has ruined hundreds of thousands of dollars' worth of gasoline trucks by shortening their efficient life."[40] The electric, by contrast, had "all the speed consistent with economy," yet could not be driven "thirty miles per hour or so at the will of an irresponsible driver when he is beyond your immediate jurisdiction."[41] Because electric vehicles were not geared for high-speed travel, not even a daredevil driver could

Figure 4.2

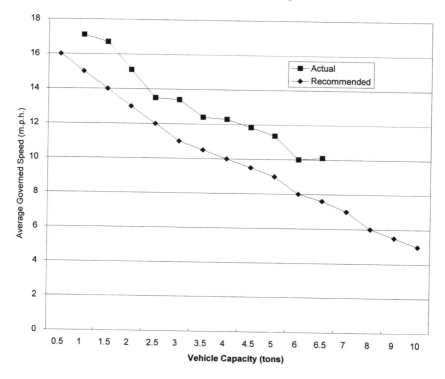

FIGURE 4.2. Recommended versus actual governed speed of commercial vehicles, 1912. *Source: "Speed Governors on Commercial Vehicles," Commercial Vehicle 8 (April 1913): 28–29.*

make the truck exceed its designated speed limit. Ironically, then, the technological limits of the electric vehicle made it more appealing to fleet owners concerned about potentially abusive drivers.[42]

Owners and makers of gasoline trucks first responded to the problem of overspeeding by installing governors that limited the speed of the engine or the truck. In contrast to the slow pace of electric starter adoption, governors spread quickly throughout the gasoline commercial vehicle market. By 1913 a simple engine limiter had been developed: a butterfly valve restricted the flow of gasoline to the motor whenever engine speed exceeded a set value. Figure 4.2 shows the governed speed recommended by the Commercial Vehicle Manufacturers Convention in March 1912 and the actual average governed speed for different-sized trucks. Although it was in the interest of each manufacturer to honor the recommended speed limits, a slightly faster vehicle was easier to sell; so a vehicle maker could gain strategic advantage by setting the governor one mile per hour faster than all the competitors' models. Baker Motor Vehicle representative George Kelly framed this dilemma nicely: "While speed is the curse of the motor truck business, it is always necessary to prove speed to persuade the horse owner that he should change his delivery. . . . There is a point up to which speed is economical and beyond which it is a great expense."[43] Moreover, the presence of a governor did not always solve the speed problem. Drivers often disabled them and justified such tampering by claiming that the truck had been stuck in mud or on a hill and needed more power in low gear than the governor would allow. Other drivers simply admitted that they had deactivated the governor because they did not like it.[44] The Duplex Engine Governor Company of Brooklyn responded with a more sophisticated governor that promised to permit full power in low gear, but the product never captured a significant share of the governor market.[45] Some manufacturers locked the governor unit in such a way that only the vehicle owner could adjust the maximum speed; others went even further, sealing the unit completely and claiming that any sign of tampering invalidated the customer's warranty.[46] Despite the broad diffusion of governors, attempts to restrict the speed of gasoline trucks were unable to solve the driver problem.

Ultimately, governors and other technologies for monitoring and surveillance (for example, a special speedometer that reported the maximum speed attained during the course of the day) failed to solve the driver problem.[47] Owners gradually abandoned technological restraints and turned instead to incentives and organizational solutions that linked drivers to a single truck and rewarded those whose trucks performed better than average. As early as 1907, *Power Wagon* had underscored the value of a prudent driver and suggested that "the surest way to encourage careful driving of motor wagons would be to offer a bonus or premium for maximum vehicle mileage with minimum repair charges."[48] One truck club created a "mutual supervision department" that distributed postcards to volunteers who were encouraged to report observed abuses to the club, which then passed the reports along to the owners of the vehicles. Thus, long before owners decorated their vehicles with toll-free phone numbers and bumper stickers asking "How am I driving?" fleet operators hoped to use the public to monitor and report instances of abusive or hazardous driving.[49] In practice these organizational solutions proved more effective than technological constraints, especially in the postwar period.[50]

SETTING BOUNDARIES AMONG SEPARATE SPHERES

By 1910 the idea of appropriate spheres was firmly established within the commercial motor vehicle industry. Under the headline "Do Not Make Civil War," a 1912 editorial in *Commercial Vehicle* started with the assumption that "the electric and the gasoline truck each have a separate and different field." These fields occasionally overlapped, according to the editorial, but only when personal prejudice rather than efficiency determined the choice. Ultimately, it was ridiculous to compare the two fields: "All comparisons are odious." From the journal's perspective, it was simply a question of educating consumers about the exact field of each motive power.[51] A survey of 1913 trucks was similarly matter-of-fact: "The fields of the electric and gasoline truck are becoming more definitely and universally recognized as separate and thus competition between the two is becoming more and more a thing of the past. . . . [Mixed] systems are becoming fashionable."[52] No less an authority than the manager of the General Motor Truck Company observed in 1914 that "today more than ever before the transportation problem is not to be settled by an off-hand declaration, either in favor of gasoline or electric trucks."[53] Indeed, General Motors had invested in separate spheres by purchasing the Landsen electric truck company in 1912.[54] The company continued to offer a line of electric trucks through the 1916 model year.[55] Thus, in the early 1910s the commercial vehicle industry accepted the principle of separate spheres at least in theory.

In practice, however, it was difficult, if not impossible, to establish and maintain the boundaries among the different service fields. The continuing role of horses, for instance, posed one important question: were there to be two or three spheres? *Commercial Vehicle* counseled against dismissing the horse and defended as "the simple truth" the idea that each motive power, the horse included, had its own appropriate market niche.[56] The relative price advantage and flexibility of single horses offered one justification for their continued use. They were easily and quickly bought and sold, thereby allowing companies to use a few more or less depending on seasonal variations in demand. This flexibility, however, was predicated on the firm maintaining a stable. If the stable were completely converted to a motor vehicle garage, the same flexibility could be obtained by keeping several spare batteries in the storeroom or sending a gas truck out for an extra circuit.[57] Even the most committed supporters of separate spheres must, therefore, have accepted the inevitable prospect of only two different spheres: one for gasoline and one for electricity.

Regardless of whether there were two or three spheres, vehicle operators were faced with the difficult task of deciding how to dispatch vehicles for specific duties. To assist the fleet owners, a number of different heuristics were proposed to help establish the boundaries among the separate spheres. Where one stood depended, of course, on where one sat. EVAA president Frank Smith, for instance, claimed that electric vehicles could meet 98 percent of all transportation needs.[58] Noted electrical engineer Charles P. Steinmetz, at a talk before NELA in June 1914, suggested that an electric vehicle capable of traveling thirty miles on a single charge could satisfy 90 percent of all transportation requirements.[59] Warehouse owner Walter C. Reid was more modest still, proposing a twenty-mile range encompassing 80 percent of all city work.[60] MIT researchers Harold Pender and H. F. Thomson did not estimate the percentage of total commercial service that could be provided by electrics but did suggest that forty-five miles per day was the maximum effective range of any electric commercial vehicle.[61] The National Bureau of Standards concluded that electrics were most economical within a ten-mile service radius.[62] General Vehicle Company sales representative Day Baker tried to clarify matters by offering two guidelines: "Never try to make deliveries with gasoline trucks if they *can* be made with an electric" and "After leaving the delivery room, don't stop the gasoline truck until beyond the fifteen-mile limit."[63] Perhaps William Blood, an ex-president of the EVAA, summed up the situation best when he opined that the "car for the middle field of moderate distances with frequent stops is the electric."[64] The experts, even within the electric vehicle industry, did not speak with a single voice.

Definitions of service compounded the problem of deciding which kind of truck to use. Should trucks and horses be compared according to total costs? Certainly not. Trucks cost more than horses, yet they also did more work. The ton-mile emerged as the favored rubric but not without a struggle. Delivery providers, for instance, were more concerned about the number of packages delivered than about the weight of the packages. Using total costs per package, electric trucks often outperformed the slower horse-drawn wagons and more cumbersome gasoline trucks.[65] A 1924 report by Edward E. La Schum, superintendent of the American Railway Express Company, conceded that gasoline trucks were always cheaper to operate than electrics on a ton-mile basis. He argued, however, that it was the cost per route, not the cost per mile or per truck, that justified the choice of one technology over another.[66] Other operators favored a measurement system that accounted for time as well as weight and distance; however, their creation—the ton-mile per hour—did not find broad acceptance.[67] Even operators who did choose to measure service using ton-miles carried, disagreed over how to categorize trips that involved multiple loading and unloading points. The commercial ton-mile was proposed to distinguish commercial service from absolute ton-miles hauled.[68] The details of these different accounting measures are not important; but the way service was defined and measured did affect assessments of efficiency and value, and these items could be problematic. One study, for instance, tried to incorporate standing time, speed, and ton-miles carried into a single score, but the complexity of the scheme limited its appeal.[69] Because different spheres provided different services, competing definitions of service were unable to help operators establish separate spheres *ex ante;* instead, measurement systems were ultimately used only to monitor relative efficiency after a given type of vehicle or service plan had been adopted.

Without a clear set of guidelines to help fleet owners establish separate spheres, companies improvised their own rules of thumb. A 1915 study of the delivery patterns prevailing in six cities suggested the range of practices that had emerged.[70] The six—Boston, Hartford, Cincinnati, Louisville, Kansas City, and Los Angeles—presented widely differing geographical settings. Maps and data summarizing delivery zones in the cities presented invite several observations (see figure 4.3). First, separate spheres were indeed in evidence; however, the two most identifiable zones were those for horse wagons and gasoline trucks, not electric and gasoline trucks. The sphere of the electric, shown only in Boston and Kansas City, was no larger than that of the horse in the other cities. Moreover, although local geographic conditions influenced the maximum range of economical delivery by gasoline truck, those circumstances did not alter the relative role played by each of the motive powers. A companion report on department store deliveries in Pittsburgh illustrated an alternative way of mapping separate spheres (see figure 4.4). The Pittsburgh merchants surveyed used gasoline trucks almost exclusively (164 out of 171) to deliver to three zones of seven-, fifteen-, and thirty-five-mile radii from downtown. Rather than use different types of delivery vehicles, the different zones were covered more or less frequently according to distance. The seven-mile zone was served four times daily, the fifteen-mile zone twice, and the thirty-five-mile zone only once. In Pittsburgh, therefore, the gasoline truck seemed to have established universal service.[71] Figures 4.3 and 4.4 also offer a dramatic illustration of the mathematics of surface area that confronted the electric truck. The gasoline truck covered more than twenty times the area served by a horse team, while the electric could only extend the horse-based service area by a factor of 3.5. In the six cities studied, electric vehicles did not offer much more than horses, nor could they approach the potential coverage provided by gasoline.

How else were appropriate spheres employed? Journal reports between 1912 and 1915 recount many instances in which observers were allowed to ride along with truck drivers. Examining forty-five such reports (fourteen electric vehicles and thirty-two gasoline trucks), we see that electric trucks averaged just under half as many miles per day (21.8) as the gasoline vehicles (47.0). Five out of eleven (45.5 percent) of the electric trucks were light delivery wagons, compared to only 15.6 percent of the gasoline trucks. In general, the electric trucks averaged slightly slower speeds than did the gasoline ones.[72] According to this sample, electric vehicles could cover a 373–square-mile service area (10.9–mile radius), while the gas trucks, with an operating radius of 23.5 miles, could cover an area of 1,735 square miles. As was the case with the six-city study just discussed, the sphere of the electric vehicle was considerably smaller than that of the gasoline truck.

If gasoline trucks were better suited than electric vehicles to longer delivery routes, another question naturally follows: were electric vehicles actually better at covering short delivery routes? Because all three power sources were in principle capable of providing short-route service, the answer turned on the issue of costs. Complete operating costs, however, were excruciatingly difficult to calculate. The trade literature was littered with reports claiming to document the relative advantages of one or the other choice of motive power; but most of these studies were incomplete, reporting detailed costs for only one motive power or failing to account for certain critical categories of expense. Other surveys pitted apples against oranges— for example, comparing horse-drawn vehicles in one line of work with electric or gasoline vehicles in another. Still others would take one limited set of data and extrapolate cost and performance based on that data to an entire fleet or region.

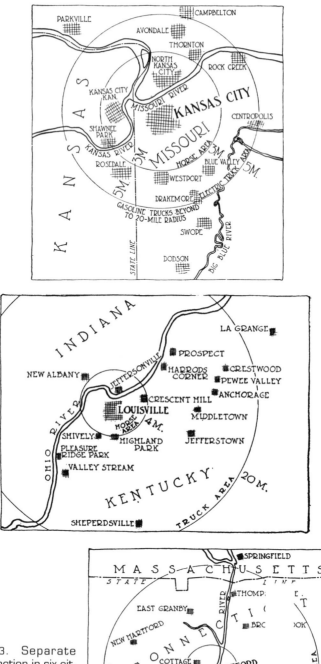
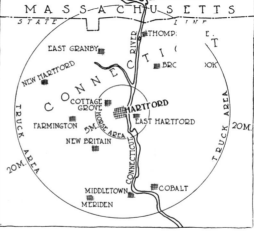
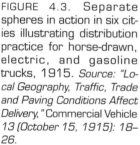

FIGURE 4.3. Separate spheres in action in six cities illustrating distribution practice for horse-drawn, electric, and gasoline trucks, 1915. *Source: "Local Geography, Traffic, Trade and Paving Conditions Affect Delivery,"* Commercial Vehicle *13 (October 15, 1915): 18–26.*

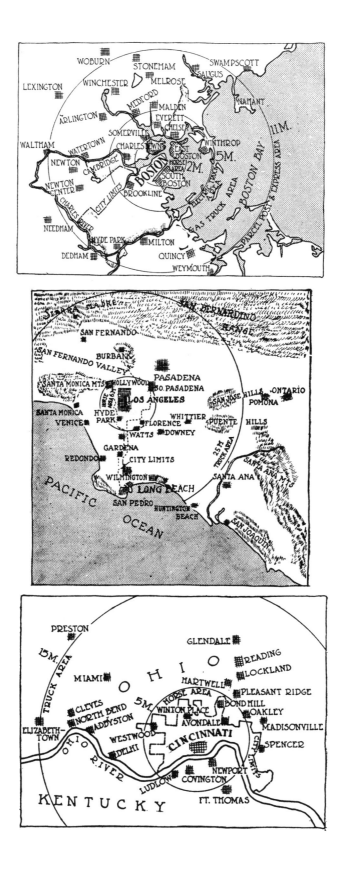

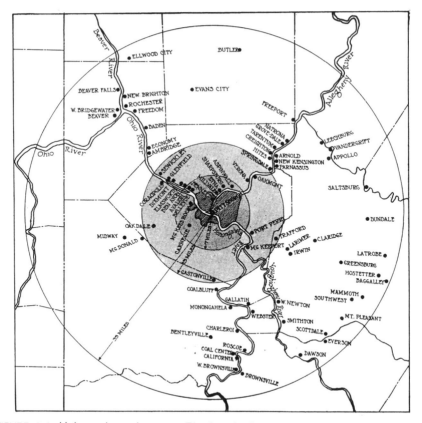

FIGURE 4.4. Universal service at a Pittsburgh department store using gasoline delivery trucks, 1914. *Source: "Trucks Give Pittsburgh Department Stores Better Service at Less Cost," Commercial Vehicle 11 (August 15, 1914): 5–8.*

Despite these evident shortcomings, several reliable and interesting studies were produced. Three representative reports, spanning more than a decade and using different approaches to evaluate costs, are particularly instructive. All three concluded that, within a specific range of operation, the electric vehicle was the most efficient transportation service option. The first study, prepared by Chicago Edison in 1909 and published the following year in *Electrical World,* concluded that for city delivery in Chicago electric vehicles cost between $0.19 and $0.47 per mile, while horse-drawn wagons in comparable service cost between $0.21 and $0.27 per mile. As table 4.1 shows, if an electric truck simply replaced a horse-drawn wagon, operating costs per mile increased 70 percent due to higher capital, maintenance, and labor costs. But when average service mileage rose to twenty-seven miles per day, the electric vehicle cost 5 percent less to operate than did the horse wagon. When Chicago Edison extended

Table 4.1 *Relative costs of delivery by electric vehicle, Commonwealth Edison, Chicago, 1909, illustrating benefits of combining motorization and organizational reform*

	ELECTRIC			HORSE	
	15 Miles/Day No Helper	27 Miles/Day No Helper	43 Miles/Day with Helper	15 Miles/Day No Helper	30 Miles/Day with Helper
Cost per Mile	$0.47	$0.26	$0.19	$0.27	$0.21
Relative to traditional practice	170%	95%	68%	100%	76%

Source: "Central Station Experience with Electric Vehicles," *Electrical World* 55 (January 13, 1910): 110–13.

the route to forty-three miles per day by adding a package boy to assist the driver with neighborhood deliveries, savings increased to 32 percent over horse-based practice. Not surprisingly, switching horses at midday, adding a package boy, and increasing the mileage of the horse-drawn vehicle to thirty miles per day also reduced operating costs, but only by 24 percent relative to traditional methods. In short, the Chicago Edison study supported the anecdotal reports on the organization of transportation service: individually, technological and organizational innovations could yield cost reductions; but to achieve maximum cost savings, both types of changes needed to be implemented together.

The second set of studies, carried out over a three-year period by MIT researchers Pender and Thomson, was probably the most comprehensive analysis of relative delivery costs. With funding from Boston Edison and the support of the EVAA, Pender and Thomson collected data from dozens of firms engaged in a broad range of transportation-related activities. Whereas Chicago Edison focused only on its own transport needs, Pender and Thomson evaluated total costs for at least five different activities, including suburban parcel service, city delivery, and beer, coal and furniture delivery service. Again, using horse-drawn delivery with two teams as the baseline for comparison (see table 4.2), the MIT study concluded that electric vehicles were more economical than either gasoline trucks or horse wagons in all five areas and recommended that electric vehicles be used whenever they were capable of doing the job. The authors amended these recommendations to include a short-haul role for horse wagons too. By the time the researchers completed the later phases of their research in 1915, they had identified three distinct economic zones—one each

Table 4.2 Relative costs of delivery by electric vehicle, MIT final report, 1913. Horse-team data are based on changing teams at midday.

SERVICE	SUBURBAN PARCEL	CITY DELIVERY	FURNITURE	BEER	COAL
PAYLOAD	1,000 LB	1,000 LB	4,000 LB	7,000 LB	10,000 LB
Electric					
Miles/Day	37	33	32	31	28
Cost as Percent of Horse	79	93	89	80	76
Gasoline					
Miles/Day	39	34	33	33	32
Cost as Percent of Horse	88	111	108	95	86
Horse					
Miles/Day	29	24	25	22	19

Source: Harold Pender and H. F. Thomson, "Observations on Horse and Motor Trucking," *Central Station* 12 (April 1913): 330–31.

for horses (less than a two-mile radius), electric trucks (a six- to ten-mile radius), and gasoline (beyond ten miles), depending, of course, on local service conditions.[73]

The third study, prepared by American Railway Express Company transportation supervisor Edward E. La Schum, dates from 1924. American Railway Express provided delivery services in more than four hundred cities and towns nationwide. Using data on more than 18 million truck miles from ten representative cities, the study found that electric trucks were considerably cheaper to operate than gasoline ones. Unlike Chicago Edison and MIT, however, the company did not report comparative costs per mile. Although La Schum readily admitted that "the acid test of comparative costs" was the only way to select appropriate transportation technology, he nonetheless chose to report costs per truck day. In the context of multi-vehicle fleet operations, separate spheres could be justified relatively easily; there were always enough local deliveries that could be handled by electric vehicles to warrant maintaining multiple technologies. According to La Schum, 85 percent of city delivery service fell within the operating realm of the electric vehicle, and under these conditions the cost benefits were significant. The average costs for all ten cities showed that an electric truck cost 38 percent less to operate than a gasoline-powered one ($3.15 per day versus $5.07 per day; see table 4.3), leading La Schum to estimate the potential national market for light-duty electric trucks at 600,000 vehicles.[74] With several thousand vehicles in continuous service nationwide, electric vehicles likely saved the express company as much as $750,000 in annual operating costs over an all-gasoline fleet.[75]

Taken together, these studies—spanning thirteen years and ranging from a single, site-level comparison between horses and electric trucks to multi-city, multi-modal evaluations of horse, gasoline, and electric service—suggest a sphere within which electric vehicles were more economical than either horse wagons or gasoline trucks. Even in the 1920s, as the role of the horse gradually diminished, the electric truck could lay claim to its own distinct realm of economic and operational advantage. But these advantages hinged on the assumption that separate spheres existed. Yes, it was true that if a merchant maintained multiple types of prime movers, there was a delivery window within which electric trucks were superior. But the studies failed to account for organizational and geographic changes that were occurring independent of motorization. The narrow service requirements suitable for the implementation of separate spheres were changing even as the researchers were collecting their data.

CHANGING TRANSPORT NEEDS AND THE COLLAPSE OF APPROPRIATE SPHERES

Several parallel processes marked the failure of the idea of separate spheres. First, purchasers of commercial vehicles refused to accept the existence of distinct fields of operation. Despite being told that comparisons between the different vehicles were "ridiculous" and "odious," consumers insisted on making the comparisons anyway. In practice users compared different types of vehicles as if they competed head to head. In letters and in person, would-be truck owners weighed the different motive powers against each other. In one instance, a reader wrote to *Commercial Vehicle* in search of comparative operating data. The journal replied with the party line: a true comparison could not be made because the two fields were "quite distinct."[76] A Toronto truck salesman responded the following month: "Comparisons in this city are inevitable, where all the work can be done by gas truck, but . . . might be better done with electrics." In truth, comparisons were inevitable in all cities, and the Toronto sales agent noted that he had "to defend even my right to take up his [the customer's] time discussing the electric."[77] Even vehicle producers failed to honor the idea of separate spheres. Waverly Electric's Herbert

Table 4.3 Relative costs of delivery by electric vehicle, American Railway Express Company, 1922

| | CITY NUMBER | | | | | | | | | | |
	1	2	3	4	5	6	7	8	9	10	Average
Miles traveled											
Gasoline	3,467,458	1,189,254	253,972	808,039	2,284,526	716,780	678,043	318,059	288,062	250,776	
Electric	2,661,241	777,954	328,084	1,239,121	634,573	503,642	331,539	201,035	365,915	216,420	
Fuel											
Gas/Oil	$1.81	$1.92	$2.02	$1.73	$1.31	$1.48	$1.56	$1.05	$1.27	$1.96	$1.61
Charging	$0.65	$0.42	$0.42	$0.60	$0.45	$0.48	$0.48	$0.41	$0.66	$0.56	$0.51
Battery Maintenance	$0.94	$0.61	$0.19	$0.19	$0.83	$0.55	$0.12	$0.68	$0.65	$1.41	$0.62
Electric Total	$1.59	$1.03	$0.61	$0.79	$1.28	$1.03	$0.60	$1.09	$1.31	$1.97	$1.13
Garage expenses											
Gasoline	$1.69	$0.94	$1.81	$2.13	$1.49	$1.10	$0.99	$0.85	$1.34	$1.78	$1.41
Electric	$1.01	$0.67	$1.37	$1.18	$1.10	$0.96	$1.32	$1.40	$0.84	$1.70	$1.16
Chassis repair											
Gasoline	$2.81	$3.00	$2.76	$3.14	$1.87	$2.01	$2.04	$1.14	$0.23	$1.50	$2.05
Electric	$1.52	$1.35	$0.71	$0.86	$1.19	$1.27	$0.36	$0.22	$0.43	$0.75	$0.87
Total costs											
Gasoline	$6.31	$5.86	$6.59	$7.00	$4.67	$4.59	$4.59	$3.04	$2.84	$5.24	$5.07
Electric	$4.12	$3.05	$2.69	$2.83	$3.57	$3.26	$2.28	$2.71	$2.58	$4.42	$3.15
Electric as percent of gasoline	65.3	52.0	40.8	40.4	76.4	71.0	49.7	89.1	90.9	84.4	62.1

Note: All costs reported per truck per day.
Source: Edward E. La Schum, *The Electric Motor Truck: Selection of Motor Vehicle Equipment, Its Operation and Maintenance* (New York: U.P.C. Book Company, 1924), 219–22.

H. Rice admitted that there was competition between gasoline and electric vehicles, the Gould Storage Battery Company claimed that using extra batteries allowed the electric to compete with the gasoline truck, and electric vehicle manufacturers went to great lengths to tout the long-range capabilities of their vehicles (see chapter 5).[78] A 1916 editorial in *Electric Vehicles* was even more explicit, dismissing the idea "that there is no competition between the gas car and the electric" as "a mollycoddle, pacifist argument" that "ought to bring failure to its proponents."[79] Functional specialization was never a technical requirement. Rather, it was a convention that emerged from the horse-paced era that preceded motorization, and industry leaders proved unable to sustain it.

Changing expectations of the motor truck also contributed to the collapse of separate spheres. American cities experienced unprecedented growth in population and physical scale during the last decade of the nineteenth century and first two decades of the twentieth. Suburbanization predated the introduction of the commercial vehicle, but the truck allowed businesses to follow commuters into suburbia unfettered by the limited range of horses. In Los Angeles, for instance, the effective size of the delivery zone for the Broadway department store grew from 250 square miles to more than 1,700 square miles in little more than a decade.[80] By 1911 Packard gasoline trucks used by Wanamaker's store in Philadelphia were traveling 110 miles per day.[81] In 1910 the Cranford Construction Company in New York found that its Mack, Hewitt, and Reliance trucks were costing more to operate per ton-mile than its horse teams. But by 1913 service had been reorganized, average daily truck mileage had increased 55 percent to fifty-four miles per day, and relative costs had been reversed. Commercial trucking had become cheaper.[82] The war accelerated changes in the expanding performance envelope of the commercial truck, and by the fall of 1918 a short-haul delivery truck might cover 150 miles on a given trip.[83] Even vehicle speed, long considered a liability to both truck and merchandise, was gradually becoming an asset as delivery companies sought, in E. J. Bartlett's words, "speed consistent with economy."

BATTERY SERVICE: TOO LITTLE, TOO LATE

The battery exchange service developed by the Hartford Electric Light Company was one of the most creative responses to the challenge of maintaining the separate electric sphere. Given the proximity of the light company to the Electric Vehicle Company (see chapter 2), it is not surprising that the central station experimented with using electric vehicles for its own service needs. As early as 1901, the company had purchased an electric runabout from the EVC in lieu of hiring an assistant to the general manager. According to President Austin C. Dunham, a firm believer in the need to both measure and market the benefits of electricity, the vehicle paid for itself within eight months, and the appointment of an assistant manager was "indefinitely postponed."[84] A second vehicle was added soon thereafter to enable the "head repair man" to quickly respond to emergency calls.[85] By 1905 the company reportedly operated a small garage to maintain its electric vehicle fleet.[86] A 1910 profile in *Electrical World* described Hartford as one of the "most progressive of American central stations."[87]

Like other progressive central stations, Hartford Electric had become an agent for electric vehicles, choosing to represent the General Vehicle Company line of trucks.[88] According to P. D. Wagoner, one of the initial organizers of the Hartford plan, "there were a few electric commercial vehicles purchased by merchants in Hartford and then the increase seemed to absolutely stop. For a considerable time there were practically no sales of electric vehicles in Hartford."[89] Soon thereafter, in an effort to expand the market for electric vehicles, the company approached the Electric Storage Battery Company and the General Electric Company, parent of the General Vehicle Company (GVC), and proposed to offer an integrated electric vehicle service. With GeVeCo battery service (named after the GVC and launched in earnest in February 1912), a consumer purchased a GVC vehicle without a battery; electricity was purchased from the utility through an exchangeable battery. The owner paid a variable per-mile charge and a modest monthly service fee to cover maintenance and storage of the truck. Both monthly and mileage charges varied with the size of the truck, and discounts were offered on the per-mile rates to encourage operators to make extensive use of their vehicles. Whenever a battery began to run low, the driver could pull into the electric company's garage and receive a fresh one, guaranteed to deliver at least 80 percent of rated capacity. The vehicles and batteries had been specially modified to facilitate rapid battery exchange.[90] Within the Hartford Electric Light Company service area, vehicle range was effectively unlimited.

The plan to exchange exhausted batteries for fresh ones and thereby expand the effective operating range of electric vehicles was the culmination of a series of earlier initiatives. The idea of battery exchange was nearly as old as the electric vehicle itself; electric vehicle pioneers such as Andrew Riker and Clinton Woods had suggested battery exchange as early as 1896.[91] L. R. Wallis, a station manager from Woburn, Massachusetts, wrote a letter to *Electrical World* in 1900 pleading that "some parent company . . . should own and control the battery sets." Wallis went on to compare the development of the electric vehicle to spread of the pianola: "the pianola has only become a useful addition to the piano since the circulating library of music has been established." For Wallis, freshly charged batteries available "at every lighting and street railroad generating station" would be like the "circulating library of music" for the pianola, a complementary technology that would increase the value of the primary technology to its users.[92] And lest one think that these early proposals had been forgotten, EVAA president William H. Blood, in his first address to the organization in October 1910, spoke of "looking forward . . . to the time of universal exchange of batteries, considering the battery, if you please, simply as a carrier of energy in much the same manner that we look upon the bottle as the receptacle to hold the milk."[93] Arguments about cooperation and providing dependable maintenance and garage service led inevitably to the concept of battery exchange. But most of all, battery exchange offered a potential solution to the persistent problems associated with using a single battery for a single vehicle. Battery service severed vehicle from battery once and for all. Implementing battery exchange amounted, in effect, to admitting that the battery was a critical barrier to the further spread of the electric vehicle. Electric vehicle supporters had mounted many rhetorical defenses: batteries had improved; range and speed were unimportant; electric vehicles were cheaper to operate in the long run; electric vehicles had their own separate sphere. Battery exchange, however, was an admission that all these rhetorical strategies—if not, in fact, wrong—had been ineffective.

On the surface, the battery service system solved many of the lingering problems associated with commercial electric vehicles. First, it allowed small merchants to purchase and use electric trucks. Previously, only operators of large vehicle fleets could afford to take full advantage of the electric vehicle opportunity because of the high unit costs of batteries, charging equipment, and experienced maintenance personnel. Using the average battery-to-vehicle ratio of the Hartford service as a measure of efficiency (eighty-seven batteries for sixty-four vehicles, or 1.36 batteries per vehicle), a local grocer would have needed to employ at least three vehicles to justify the purchase of a fourth battery. Anecdotal reports from users in 1907 had indeed suggested that three vehicles were the minimum necessary to justify the investment.[94] The minimum efficient scale necessary to employ a skilled garage supervisor was larger still, probably on the order of ten to twelve vehicles.[95] In moderate-sized cities such as Hartford, limiting the market of potential purchasers to those willing to buy three or more vehicles placed the electric vehicle at a severe disadvantage. Only a small number of fleet operators could efficiently use three vehicles, let alone twelve or more. By shifting responsibility for charging and maintenance from the vehicle owner to the service provider— in this case, the central station garage—battery exchange allowed an entirely new class of vehicle purchasers to consider the electric option. Although there are no detailed data on the distribution of fleet size among battery service customers, it is significant that 63 percent of the GeVeCo battery system customers purchased electric trucks to replace gasoline trucks and that other reports indicated success among new motor vehicle users. A drug store, a dry goods house, a bakery, and even a distributor of horse feed all bought their first trucks through the battery service plan.[96] During a time when most electric vehicles were being sold to large fleets as reorders, battery service lured a number of first-time buyers to select electric trucks.

In addition to expanding the market of potential electric truck customers, battery service also increased the appeal of electric vehicles for the existing base of users. Battery service was trouble-proof; the vehicle owner no longer had to contend with the array of potential problems that might have caused a battery to malfunction. The service resolved the nagging uncertainties surrounding the use of commercial vehicles. From the largest fleet operator to the smallest corner grocer, every battery service subscriber could determine exactly how much its transport service was costing by simply looking at the odometer. And because customers bought trucks without batteries, battery service lowered the initial purchase price of the vehicles.[97] A 1916 survey article estimated that battery service reduced the initial purchase price by as much as 38 percent.[98] Thus, although the EVAA had identified many of the institutional barriers to the expansion of the electric vehicle system, battery service—by actively integrating the various components of the electric vehicle system into a transparent, seamless whole—cut the Gordian knot and allowed the electric vehicle to deliver on its long-standing promise of simplicity and reliability.

Nevertheless, some institutional barriers remained beyond the reach of even battery service. An exchangeable battery did not liberate the vehicle operator from all range limitations. The longest daily run reported with a single vehicle was eighty-eight miles, less than might be achieved with a gasoline truck; and the driver still had to be aware at all times of the distance from the battery exchange station.[99] Because the Hartford Electric Light Company never established remote battery exchange stations, battery service only increased the daily mileage of electric trucks for local, short-haul delivery. The vehicles could cover more ground, but their absolute range was still limited by the capacity of a single battery. Before 1910 commercial vehicles needed only to provide short-haul service; but by the time battery service was introduced, longer hauls were already on the horizon. Nor did battery service address the driver problem and the general mechanical uncertainties associated with any complex piece of machinery. Trucks still broke down and were subject to the whim of the operator. In response, Hartford Electric Light offered an additional "mechanical maintenance contract" whereby the vehicle owner paid the company garage a fixed annual sum for repairs. If the truck required less than the prepaid amount of mechanical service, the surplus was shared as follows: "60 per cent returned to the customer; 20 per cent kept by the light company; and, provided the owner approves, 20 per cent to the driver of the truck as a bonus." Few owners, however, were convinced of the value of this extra service; and as of April 1917, only fifteen out of eighty-eight battery service vehicles were enrolled in the extended maintenance plan.[100]

Finally, battery service ran afoul of those vehicle makers who had not been chosen to participate in the battery exchange system. In Hartford and Spokane, two cities in which battery service was first tested, there were few of these "angry orphans"; but when Boston Edison, General Vehicle, and the Electric Storage Battery Company announced their intention to offer battery service in Boston in the spring of 1915, the rest of the local electric vehicle industry was incensed.[101] An ad hoc committee formed under the leadership of M. E. Brackett, the Boston branch manager for the General Motors Truck Company, with representatives from the United States Light and Heating Company, the Edison Storage Battery Company, and a handful of vehicle producers. Committee members argued that their products had been excluded from the battery service experiment and that protocols for battery exchange could be established to allow all local suppliers to participate.[102] A *Commercial Vehicle* editorial endorsing the battery service plan concluded that "strife in the ranks . . . may succeed more than the long-standing harmony of inaction, in galvanizing some life into the electric vehicle business."[103] The complaints of the orphans underscored the internal tensions that had plagued the EVAA standardization committee. As noted in chapter 3, modest standardization allowed technical compatibility among the various manufacturers. Yet to use battery exchange to significantly expand the scope of action of the electric vehicle, supporters would have had to establish national standards for storage batteries and battery compartments. Day Baker urged the Electric Vehicle Section of NELA to do just that: "The association now has within its power, as the real arbiter of the destiny of the electric truck, to establish battery compartment standards. These must permit the interchange of any make of battery which is constructed on lines and with characteristics approved by the association."[104] But as we have seen, Baker was mistaken: the industry association did not have the power to enforce such sweeping standards, and few companies were willing to surrender existing markets and proprietary technologies for the uncertain prospects of a thoroughly standardized, undifferentiated electric truck. As a result, industry-wide battery compartment standards were never adopted.

Instead, different cities developed local variations of battery service. In Boston, where the opponents to full battery exchange were well organized, service never expanded beyond offering maintenance contracts for batteries. Similarly, in Chicago the Commonwealth Edison Company organized a service that covered all costs of battery charging and maintenance but did not offer actual exchange of batteries. Four leading vehicle makers—Walker, Ward, General Vehicle, and Couple-Gear—participated, but each vehicle was still linked to a single battery.[105] Nevertheless, due to the continued popularity of the electric passenger vehicle in Chicago, private garage operator Harry Salvat organized a proprietary battery exchange system for owners of Milburn electric passenger vehicles. Starting in late 1917, purchasers of a redesigned Milburn Light Electric could buy the car for $1,485 without battery (a reduction in initial cost of approximately two hundred dollars) and sign up for the "Exchange Power System" for fifteen dollars per month plus one dollar per change of battery. This plan offered subscribers a further advantage not available under the Hartford system: batteries could be exchanged at any one of five stations "strategically located in Chicago and its suburbs."[106]

Given that battery service solved some but not all of the problems blocking the spread of electric commercial vehicles, how did it work in practice? Did it sustain or even expand the separate sphere of the electric vehicle by allowing small firms to use electric trucks to provide effective commercial service? Apparently so. D. C. Perkins, owner of a Hartford delivery company, purchased his first electric vehicle on battery service in 1912 and by 1917 had expanded his fleet to include seven electric trucks. He claimed that the vehicles earned a 40 percent return on his investment and allowed him to profitably cover "long runs" that "were formerly done at a loss" by horses. And although Perkins's drivers initially resisted the electric trucks, believing "that it was a scheme for helping the boss," they, too, came to prefer the motor vehicles, "principally because they [the drivers] don't have to take care of horses after they get through work and on holidays."[107] Even following the hardships of World War I, when the Hartford Electric Light Company offered refunds to potentially dissatisfied customers, none accepted the offer. Despite the temporary inconveniences resulting from the war, battery service subscribers preferred to keep their vehicles enrolled.[108]

In addition to these anecdotal accounts, limited panel data are available from the Hartford experiment with battery service. These reports, summarized in figure 4.5, suggest that battery service was an effective response to the uncertainties of range and cost that had plagued electric commercial vehicles for more than ten years. From December 1911 to January 1919, the total number of battery service subscribers grew from fourteen to ninety-two, where it remained until at least 1924. Over the life of the service, average mileage per vehicle reached thirty miles per day; and in specific instances, owners reported average mileage as high as forty-nine miles per day.[109] Although (based on the preceding analysis of vehicle service routes) these daily figures did not allow the electric trucks to compete with gasoline truck operators intent on expanding their daily service range beyond one hundred miles, the range of the battery service vehicles was considerably greater than that of a typical horse team. American entry into World War I, which led to labor shortages and reductions in transport service, gradually pushed down average daily mileage;

Figure 4.5

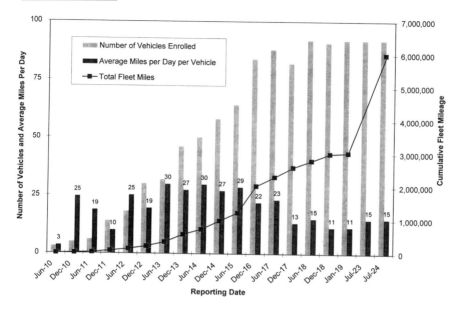

FIGURE 4.5. Number of vehicles, average daily mileage, and cumulative mileage of electric vehicles enrolled on the Hartford Battery service system, Hartford, Connecticut, 1910–24. *Source: "Developments in Electric Vehicle Trucks at Hartford, Conn.,"* Electrical World *73 (April 19, 1919): 795.*

but by the end of the decade battery service vehicles in Hartford had covered more than 3 million miles, and by 1924 total mileage exceeded 6 million.

Figure 4.6 shows the number of times batteries were actually exchanged and provides another perspective on the accomplishments of the Hartford service. Lacking detailed information about how often specific users swapped batteries, we cannot draw firm conclusions about the typical vehicle; but the data show that both energy consumption (measured by number of exchanges per service day) and number of stalled vehicles were higher in winter than in summer. This pattern is consistent with expectations: winter roads were often heavy with snow and ice, and less daylight surely required more on-board lighting.[110] Even during the winter, however, the number of exchanges never exceeded the number of vehicle days. That is, in the month with the highest number of exchanges per service day (December 1915), 1,188 vehicle days required only 1,152 battery exchanges. Although customers on battery service had the option of using more than one battery per day to extend service range, on any given day the average truck owner did not require more than a single

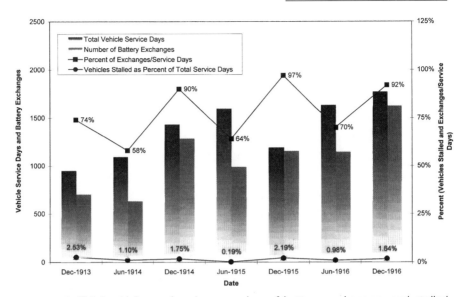

FIGURE 4.6. Total vehicle service days, number of battery exchanges, and stalled vehicles reported, Hartford battery service system fleet, Hartford, Connecticut, 1913–16. At no time did the number of battery exchanges exceed the number of vehicle service days; that is, on average, on a given day, the electric trucks enrolled in the system did not need more range than was available from a single battery. *Source: "Battery Exchange Service at Hartford," Electrical World 69 (April 28, 1917): 792.*

charge. And during the summer a truck on average seemed to require a fresh battery only four times a week.[111] Apparently, maintenance, dependability, and (as D. C. Perkins's mentioned) the potential to make occasional longer runs were the primary issues for battery service customers. In Hartford daily range was simply not a barrier for users of the electric truck.

Several minor adjustments to the battery exchange system suggest both the appeal of the program and some of the reasons for its ultimate undoing. On the one hand, by 1916 at least two hundred vehicles were enrolled in battery service programs in ten cities nationwide, and (as noted) the Hartford service survived well into the 1920s.[112] But by 1924 the Hartford Electric Light Company was no longer the primary service provider. Willis Thayer, long-time manager of battery exchange at Hartford Electric, had become president of the Electric Transportation Company, an independent provider of battery service for electric vehicles.[113] Similarly, in Chicago in 1921 Commonwealth Edison continued its battery maintenance service but resisted battery exchange. A private company rented exchangeable batteries for 220 commercial trucks and 800 passenger cars.[114] The fact that battery rental services in Hartford and Chicago were both sold off by their initial owners suggests that the services were not sufficiently profitable to justify the heavy overhead costs of maintaining a battery service station. In fact, before Willis Thayer established his independent company, he explicitly recognized the importance of distributing these costs across a large number of subscribers. In addition to simplifying the rate structure for battery service, he took the further step of offering an annual rebate for every subscriber based on the total number of vehicles enrolled. If more than fifty vehicles were enrolled on battery service, all subscribers received a 2 percent rebate of the yearly fixed charges, a figure that increased by 0.5 percent for every additional ten vehicles subscribed.[115] Thayer also offered traditional discounts to firms that purchased multiple vehicles, but the fleet-size bonus was unique in that it rewarded individual users based on the size of the group. In economic terms, Thayer recognized the network benefits (i.e., externalities) associated with having more trucks enrolled in his service. The more vehicles he maintained, the fewer extra batteries he needed per subscriber.

Battery exchange service was a creative solution to the problems associated with secondary storage batteries; and at least in Hartford, even after the end of the GeVeCo system for electric trucks, battery rental remained popular among gasoline truck owners for starting, lighting, and ignition batteries. Had battery service been instituted a decade earlier, before the gasoline passenger vehicle had stabilized, the system might have been able to expand using exchange substations to increase daily operating range, thus providing sufficient network benefits to sustain and expand the market for electric trucking.

THE IMPACT OF WORLD WAR I

The outbreak of hostilities in Europe in 1914 transformed the domestic truck industry. The European belligerents had recognized the military relevance of the motor truck several years earlier and had established parallel programs to subsidize buyers of commercial vehicles. As early as 1908, American observers at a French truck trial had remarked on the extent to which "military opinion was severely felt" in the judging.[116] In France, Britain, Germany, and later Russia, prospective truck purchasers received up to $1,200 from the government if they selected an approved truck, maintained it to military specifications, and were willing to surrender it to the state in case of war.[117] Not only did these subsidy programs tilt European markets away from electric and other nonstandard vehicle designs, but military subsidies ensured that the peacetime truck was also a military technology. Speed and range were accorded greater importance than commercial service alone would have required. War needs, however, rapidly exceeded the supply of subsidized trucks, and the Allies soon turned to the American truck industry.

In contrast to the Europeans, the American military establishment failed to incorporate trucks into war planning. In 1906 the Quartermaster Corps had invited domestic truck makers to submit proposals to supply a military convoy on exercise, but only one company responded; and the $92.50 per day price tag convinced the War Department that economical military trucks were still many years away.[118] In general, before World War I the public sector did not play an active role in the truck market, although the federal government owned several hundred trucks for use by various departments and for transportation at naval shipyards. A *Commercial Vehicle* editorial in September 1914 anticipated many of the changes the war would bring: "Motorized warfare will modernize transportation."[119] Truck exports expanded dramatically in 1914 and 1915 as American manufacturers rushed vehicles to the European theater. Following the American military incursion into Mexico in pursuit of Pancho Villa and the siege of Verdun, the American military finally swung into action.[120] In 1916, however, the commercial vehicle industry, by virtue of eighteen months of war sales, knew more about military trucks than did the Quartermaster Corps.[121] Discussions aimed at creating a standard war truck were initiated through the Society of Automobile Engineers; and by the time the United States entered the war in 1917, several standard designs were already on the drawing board.[122]

The dramatic role of motor trucks in the conduct of the Great War reinforced and accelerated the standardization of the commercial peacetime truck. First, although by 1914 there was already little doubt that gasoline trucks would become the dominant technology for suburban and long-range commercial service, the war pushed electric trucks into an increasingly marginal and inferior market position. Comparing the American and European experiences, one important difference stands out. Fuel prices in Europe, already higher before the outbreak of hostilities, increased more than did fuel costs in the

United States. Moreover, acute gasoline shortages limited general availability for civilian uses, thereby creating an umbrella that sheltered European electric vehicle fleets and preserved a measure of separate spheres. English war purchases, for example, included a number of electric delivery trucks used to replace horses and gasoline trucks that had been requisitioned for war duty. Some British firms actually imported American electric trucks because they knew that the vehicles would not be requisitioned.[123] In the United States, however, where gasoline prices never rose high enough to create a lasting cost advantage for urban electric trucks, the dominant effect of the war was to increase scale of production and lower prices of American gasoline trucks. By 1919 electric trucks accounted for less than 1 percent of the total number of commercial vehicles produced in the United States, down from 11 percent in 1909.[124]

Wartime pressure on the national railroad infrastructure and the devastating coal shortages of the winter of 1918 also focused attention on the need for an alternative to the railroads for long-haul transport. As more and more companies began to experiment with motor trucks for overland freight, road building assumed strategic and military importance. Forty years before the Eisenhower administration created the interstate highway system, mobilizing for war revealed the importance not just of good roads suitable for heavy trucking but of the layout of such roads to facilitate efficient haulage.[125] As one *Commercial Vehicle* editorial observed, "the war has shown the businessman how to use trucks in ways heretofore undreamed of."[126] After the war thousands of surplus trucks were distributed to state highway departments to further the construction of good, truck-ready roads.[127]

In sum, World War I left a lasting mark on the American commercial vehicle. Because electric trucks, once in service, tended to perform adequately for ten or more years, electric vehicles continued to ply the streets of American cities for many years after World War I. Well into the mid-1920s, companies that already owned and operated electric vehicles would often expand by adding more electric vehicles to their fleets. But for all practical purposes, the internal combustion–powered Class B Liberty truck had become the standard bearer. Commonwealth Edison, for instance, did not retire the last of its Walker electric vehicles—"old juice box 749"—until October 1947. But the company had stopped adding electric vehicles to its fleet in the early 1930s, when it sold its electric truck division to Yale

and Towne; and by World War II Edison operated more than eight
hundred internal combustion vehicles.[128]

THE UNIVERSAL COMMERCIAL VEHICLE

The end of World War I coincided with the emergence of internal
combustion as the universal commercial vehicle. Although many
thousands of horses remained in use in American cities and a number of merchants clung doggedly to their fleets of electric vehicles,
the cumulative effect of the war was to create new (interurban) markets for internal combustion and squeeze the alternatives into ever
smaller market niches. Following the war there was only one big
sphere left. Commercial motorization had begun as part of the
broader process of rationalization of American industrial practices
(that is, adapting horse-paced organizations to the modern world),
but in subsequent years motorization was its own justification. The
commercial vehicle had finally joined the automotive bandwagon.

In the context of the larger phenomenon of technological choice,
however, the decline of the electric truck resulted neither from the
random outbreak of war nor from the technical liabilities of the electric truck in commercial service. The war merely accelerated the process of technological selection. Rather, the collapse of the commercial
electric vehicle market resulted from the inability of the electric truck
to meet all the needs of all its potential customers. The peculiar success of the electric truck resulted from its similarity to the horse
wagon. The electric vehicle allowed merchants to expand service without challenging or destabilizing existing organizational structures. In
a short-haul transportation environment, where separate spheres were
dictated by the continued use of horses, the electric seemed to be just
different enough and therefore worthy of its own appropriate sphere.
Numerous reports proved that under appropriate conditions the electric truck was more economical to operate than its gasoline-powered
cousin. But with the gradual demise of horse-based transport, the approval of more fundamental changes within delivery systems, the
standardization of the ton-mile, the acceptance of speed and range
as necessary components of long-haul trucking, and the inability of
proponents of separate spheres to practically define the economic
boundaries between supposedly distinct fields of action, the appropriate sphere of the electric truck grew smaller and smaller. Local merchants, faced with the choice of either internal combustion or

electricity—but not both—almost invariably opted for internal combustion because only gasoline trucks could provide universal service. Had a hybrid market for passenger automobiles emerged or had battery service been introduced a decade earlier, perhaps the situation would have unfolded differently. As it was, however, the electric vehicle was slowly relegated to increasingly narrow fields of action—industrial trucks (moving freight inside warehouses and factories) and personal mobility (motorized wheelchairs and golf carts)—until its rediscovery in the early 1960s.

5

INFRASTRUCTURE, AUTOMOBILE TOURING, AND THE DYNAMICS OF AUTOMOTIVE SYSTEMS CHOICE

If a man wishes for cheap electric current for his electromobile, he can of course install an oil engine to run a generator to charge a storage battery to run a motor on his carriage to operate it as far as the next charging point. If, however, he sees no beauty in the "house that Jack built" method of applying energy, he can give up his "pipe dream" of electrical traction, take his oil engine and put it on the carriage, and go where he likes and as far as he likes, and not be a slave to a wire.

Albert Clough, *Horseless Age,* August 1902[1]

Col. Bailey says, "What is the use
Of touring without any juice?
    With my own private plant
    Don't tell me I can't
Burn the dust off the road like the deuce.

Anonymous, Electric Motor Car Club of Boston, 1913[2]

It is futile to answer that once there are a multitude of electric cars in use service stations will naturally spring up—for you cannot get the multitude without them. Lack of convenient charging facilities is the main reason that the electric vehicle, with all its manifest advantages, lags behind its opportunities.

Editorial, *Electrical World,* August 1920[3]

Along with a small potted plant, the preceding limerick was presented to Colonel E.W.M. Bailey at the holiday dinner of the Electric Motor Car Club of Boston in 1913 in recognition of his many electric touring ventures.[4] Bailey, the general manager of S. R. Bailey and Company, the manufacturer of the electric touring car known as the Bailey Bullet, had mounted several private campaigns in support of electric touring, including a 258–mile jaunt through Massachusetts and Vermont timed to coincide with a regional meeting of the National Electric Light Association in Burlington, Vermont, in September 1913.[5] Soon thereafter, on October 14, he set off from Boston in a stock model of his eponymous electric roadster. On October 31, two and half weeks later, after traveling more than 1,500 miles across New England and the upper midwest, he arrived in Chicago, host city for the fourth annual convention of the Electric Vehicle Association of America. Bailey was expecting a warm reception from the assembled delegates. After all, who would have thought it possible to travel from Boston to Chicago by electric car? That evening, he was indeed the guest of honor at the Illinois Athletic Club. Unfortunately, only a few EVAA members from the Chicago area were there to celebrate his feat; the convention had ended three days earlier, on October 28, and everyone else had gone home.[6] Bailey had misjudged how difficult the journey would be and how long it would take. He had made it from Boston to Chicago in an electric vehicle, certainly a noteworthy accomplishment, but he had also inadvertently confirmed the conventional wisdom he had set out to overturn: even as late as 1913, touring by electric vehicle was not a trivial undertaking.

As details of the journey leaked out, it became clear that Bailey had encountered two major obstacles in his attempt to demonstrate the practicality of touring by electric vehicle. First, the roads through much of the trip, especially from Buffalo to Cleveland and east from Cleveland to South Bend, were "heavy" at best and "vicious" or "atrocious" in bad weather. Although he reported no mechanical problems with his Bailey roadster, the poor roads increased his energy consumption as much as fourfold over normal city driving. High current consumption alone, said Bailey, would have been manageable, except for the second problem—the appalling lack of charging facilities. He noted the "uniform courtesy and interest" of the central station managers whom he encountered along his route, but opportunities to charge were too few and far between. In one instance, Bailey was forced to prevail upon a local sawmill operator, who started his 1,200 horsepower mill engine to give the roadster a boost.[7] In short, notwithstanding the colonel's valiant efforts, touring by electric vehicle was simply not feasible for most Americans.

In 1913, despite fifteen years of technical improvements to the mechanism of the electric vehicle, the driver was still "a slave to a wire." The wire was longer than it had been in 1902, when Albert Clough decried the "'house that Jack built' method of applying energy," and longer still than in 1897, when Hiram Percy Maxim first tried to travel from Hartford to Boston and New York by electric vehicle. But the electric vehicle driver was tethered all the same.

Shifting away from specific companies, organizations, and product markets in this chapter, we will consider a different view of separate spheres and the process of technological stabilization that brought us internal combustion. Taking Colonel Bailey's experience as a starting point, we will examine why electric touring was an issue in the first place. Why did electric touring matter? To answer this question, we will look at two major factors: first, the physical infrastructure that both permitted and constrained mobility; second, the motivation and behavior of the builders, owners, and drivers who used the infrastructure. In the end, in the specific domain of the passenger vehicle, there were important marketing benefits to touring because it enabled manufacturers to sell an attractive vision of the automotive lifestyle, regardless of whether that vision translated into reality for typical early motorists.

THE ROLE OF INFRASTRUCTURE

What infrastructure supported the motor vehicle during this period, and how did its development shape the process of technological selection that resulted in the victory of internal combustion over its electric and steam competitors? In the purely physical realm, infrastructure—the structure below—is the set of conduits that we use to move from place to place. From the animal paths of pre–Columbian America to the freeways of southern California, the structure below has always directed and constrained our ability to travel.[8] The simple etymology of the phrase "path dependence" suggests the fundamental relationship between our physical paths and the destinations we select. Without a path, a route, or an infrastructure, our mobility is severely limited. The infrastructure that prevailed during this era played a major role in the outcome of the competition among steam, gas, and electric vehicles. As Colonel Bailey discovered, two major areas of infrastructure development were crucial to the would-be touring driver: the road network and the state of fuel and service systems.

Roads

Did the state of American roads tilt the playing field for or against the adoption of the electric passenger vehicle? In the 1890s the quality of the typical intercity American roadway was, to quote historian Bruce Seely, "truly abysmal—mud holes in spring, dust traps in summer, rutted always."[9] Paving of any sort was rare; many routes were simply trampled earth intermittently overlaid with loose gravel. Almost any extended overland journey by horse-drawn vehicle was arduous, slow, and uncertain.[10] Although paved roads in most major cities and towns could support motor vehicles, getting from town to town was an adventure.

During the 1890s, before the emergence of the motor vehicle, a number of different forces coalesced around the issue of upgrading road quality. The popularity of the bicycle first drew public attention to the wretched state of American roadways. As cyclists took to the roads, many were discouraged by the impassable conditions they encountered.[11] The League of American Wheelmen (LAW), a national organization of bicycling enthusiasts underwritten by none other than Colonel Albert A. Pope, encouraged local chapters to gather information about road quality and maintenance that was then published in the *Wheelman,* the house organ of the LAW. Colonel Pope, the "father of good roads," also sponsored research on roadway design and construction at MIT.[12] The U.S. Post Office's inauguration of rural free delivery (RFD) in 1896 further dramatized the need for good roads. The public clamor for RFD allowed politicians of differing stripes to increase investments in local roads.[13] Farm interests also sought to improve farm-to-market roads, and the need for improved transport was linked to a host of problems. Supporters pointed to unmeasured costs: parishioners could not make it to church, doctors could not reach their patients, children could not attend school, all because of the lack of serviceable roads.[14]

Although this first set of efforts to upgrade American roads preceded the introduction of the automobile, the nascent motor vehicle industry took an obvious interest in the state of the American road.[15] In one camp were those who believed that the roads themselves were less important than the vehicles destined to travel them. An 1899 editorial in *Motor Age* titled "Road Improvement Fallacy" encapsulated this view: "No roads would ever have been built for horses if horses had not proved that at a pinch they could give good service without roads. Don't preach that motor vehicles depend on roads. They don't. They depend on good, suitable construction to negotiate any kind of road surface, and on perfectly reliable motors."[16] Before the emergence of the standard, mass-produced automobile, roads were simply the weak link in a transport system that was making great progress on other fronts. Bad roads, in this view, were no excuse for poor design and engineering. Sound vehicles would eventually stimulate the necessary investment in transportation infrastructure. At the other extreme were those who believed that good roads necessarily preceded the automobile. In the words of one turn-of-the-century observer, "countries exist without railroads, but none without highways."[17] In support of this view, historical geographer Arnulf Grübler has argued that smooth road surfaces existed before the advent of the automobile. Citing data on miles of paved road in the United States around 1904, Grübler concludes that the infrastructure available for early cars on a per-vehicle basis "was over a factor of 100 larger than for the present car dominated transport system." From this perspective, the development of road infrastructure "significantly *preceded* the diffusion of the automobile."[18]

The debate over which came first, the road or the car, underscores the interaction between vehicle and road surface. For instance, the cobbled street, that hallmark of nineteenth-century urban road construction, was uniquely adapted for the horse hoof. Beasts of burden could exert great force at each step without losing traction. Recognizing that hoof and cobble were natural complements and fearing that drive wheels would slip on the tops of the slick stones, the builders of experimental steam traction engines in the 1820s went so far as to construct mechanical hooves instead of drive wheels. But mechanical replication of the natural motion of the hoof proved difficult and inefficient, drive wheels slipped less than had been feared, and the clumsy mechanical hooves were quickly abandoned.[19] But cobblestones proved more problematic for the first horseless carriages. The challenge was probably greatest for the electric vehicle: the weight of the lead-acid batteries increased the kinetic shock of traveling on cobbled streets; and the batteries themselves, having been designed for stationary applications, easily shed their active material, resulting in reduced storage capacity and shortened battery life. Steam and gasoline vehicle makers also had to contend with the jarring effects of the road surface. Loose nuts and broken struts were common, and drivers were expected to carry tools to cope with such occurrences.

Road and vehicle were thus inextricably intertwined, with important implications for the development of the American automobile. In France, for example, where the national road system was both more uniform and more developed by the time of the first automobiles, the selection of internal combustion as the motive power of choice occurred quickly. By contrast, the state of American roads may well have extended the period of uncertainty and competition among steam, gasoline, and electric vehicles. As late as 1917, an article on the development of the automobile in Peoria, Illinois, admitted that the city was "not by any means an ideal place in which to promote electric vehicle sales." Nonetheless, the electric vehicle was well received in Peoria, in part because the roads connecting Peoria with other cities were so poor: "when once a car leaves the paved streets it encounters roads so impassable *to any car* that the joy of motoring thus turns into marked discomfort. . . . It is an instance wherein bad roads help the sale of electric vehicles."[20] No vehicle—steam, gas, or electric—could have used Peoria as a base for motor touring. In general, poor roads were more conducive to the use of lighter steam and internal combustion vehicles; but in certain cases country roads were of such poor quality that motor vehicles were practically trapped within the narrow confines of paved borders, thus reducing the relative disadvantage of the electric alternative.[21]

Poor roads also produced regional isolation and permitted the development of temporary, local technological standards. Data on early automobile users from several cities illustrate these disparate patterns of standardization. In 1902 in Portland, Maine, steam was the initial technology of choice. Of the first fifty-two vehicles registered in this small New England city, forty-six, or 88 percent, were powered by steam; four were internal combustion (8 percent); and two were electric (4 percent).[22] In Cleveland, despite the presence of the Baker Motor Vehicle Company, a pioneering electric vehicle manufacturer, and the White Motor Company, a leading steam vehicle producer, local residents favored internal combustion from the start. Of the first one hundred vehicles registered in the city, forty-nine were powered by internal combustion, thirty-four by steam, and seventeen by electricity. By 1904, several years later, the market share of internal combustion in Cleveland had increased to 61 percent.[23] Meanwhile, in 1900 in Chicago, where commercial vehicles accounted for a larger share of early registrations, electric vehicles accounted for 65 percent of the first 303 vehicles. Internal combustion made up approximately 15 percent, with steam and hybrid cars accounting for the remainder.[24] Portland, Cleveland, and Chicago, along with the rest of the country, would soon join the internal combustion bandwagon; but at the outset local markets were isolated. The poor quality of intercity roads helped sustain the unique characteristics of these local markets.

Fuel and Charging Facilities

Poor roads were not the only impediment to touring by automobile. Access to fueling and charging facilities also influenced the ability to tour; and as the experience of Colonel Bailey demonstrated, electric vehicles were at a distinct disadvantage on this count.

Both steam and gasoline vehicles depended on the broad availability of refined petroleum. Steam cars were early examples of today's flexible-fuel vehicles: they were relatively insensitive to the exact type of fuel used to generate their steam. Boilers were generally designed to run on gasoline or lighting kerosene (the first widely available refined petroleum product), but other fuels could be used in a pinch. Coal or wood, if properly ignited and monitored, would also power a steam car for the few miles to the next town. A Norwegian correspondent in *Horseless Age* even proposed using peat briquettes.[25] Steam cars also depended on fresh, clean water from which to generate steam. Although some companies, such as White and later Doble, developed closed-circuit condensing engines that recycled waste steam, the Stanleys did not bother to add a condenser to their engine until an epidemic of hoof-and-mouth disease in New England in the spring of 1914 forced the closure of the horse watering troughs that had previously allowed Stanley owners to refill their water tanks almost anywhere.[26] Without a condenser, the range of a steam vehicle was limited to twenty-five to thirty-five miles; and over time, "briny and contaminated" water would "choke the small pipes of the machine," prompting some to call for a two-condenser design. The proposed external condenser would use water "without regard to purity or hardness," while ensuring that only pure water passed through the boiler.[27] In practice, the development of efficient condensers addressed the major limitation that the prevailing infrastructure imposed on steam automobiles.

By comparison, gasoline cars were less demanding of their water but more particular about their petroleum. Almost any water would cool an internal combustion engine. In his memoir *Horseless Carriage Days,* automotive pioneer Hiram Percy Maxim recounted with delight the experience of repairing his overheated vehicle en route to a race in Branford, Connecticut: "The water was horribly stagnant and covered with green slime, but it was wet, and that was all I cared about. . . . I cannot express the intensity of satisfaction I enjoyed as I gazed at that filthy water running into my tank. It cooled as well as the purest spring water."[28] But gasoline was another matter. The high compression developed inside the combustion chamber meant that only properly refined spirits would allow the engine to operate smoothly. A 1901 article titled "Bad Gasoline" advised motorists to travel with a hydrometer to check the degree (i.e., relative density) of gasoline before buying it. Many stores watered their gasoline or diluted it with heavier petroleum fractions, and only a hydrometer could guarantee that the fuel was sufficiently volatile for combustion.[29] In the main, however, both steam and gasoline automobiles were well served by the prevailing technological infrastructure of the late nineteenth century. Almost every rural general store stocked kerosene and gasoline for home lighting and for stationary gasoline engines (although finding a merchant who was open for business on Sunday was sometimes a challenge). In addition, early motorists could usually call on local livery stables and smithies to help with minor repairs.[30]

The general availability of kerosene, gasoline, water, and service for steam and gasoline vehicles contrasted starkly with the situation for electricity. During the 1890s urban electric grids were still in their infancy, the battle of the currents was raging, and numerous local utilities were competing to expand their service areas into adjoining regions. Towns might boast an illuminated Main Street and perhaps a store or two, but in rural areas electricity was all but unknown and would remain so for decades. For the vast expanse of rural America, the car arrived before the light bulb.[31]

The problem of finding suitable places to recharge electric vehicle batteries was compounded by several additional factors. First, batteries needed to be recharged frequently, and charging required a large block of time. In an 1897 article touting the prospects for rural touring in electric vehicles, Hiram Percy Maxim prepared the first list of cities with central lighting stations for travelers heading north or south from Hartford. It showed an adequate number of appropriately spaced central stations between Boston and New York capable of recharging electric vehicle batteries (see figure 5.1). With a range of perhaps twenty-five miles on a full charge, a driver might even be able to skip a station or two. But there was still the problem of the time it took to recharge. Maxim claimed that a short charge (ninety minutes) allowed the vehicle to travel twenty miles (almost as far as the

Town	Distance
NORTH	
Boston	7 miles
Newton	8 miles
Wellesley	7 miles
South Framingham	10 miles
Westboro	13 miles
Worcester	17 miles
South Spencer	11 miles
Warren	11 miles
Palmer	18 miles
Springfield	18 miles
Thompsonville	11 miles
Hartford	
SOUTH	
New Britain	10 miles
Meriden	10 miles
New Haven	19 miles
Naugatuck	13 miles
Bridgeport	5 miles
South Norwalk	14 miles
Stamford	8 miles
Port Chester	8 miles
Mamaroneck	5 miles
Mt. Vernon	7 miles
New York (42nd Street)	14 miles

FIGURE 5.1. Electric touring from Hartford, Connecticut, 1897, showing towns with electric generating stations willing to sell current to charge electric vehicles. *Source: Hiram Percy Maxim, "Radius of Action of Electric Motor Carriages,"* Horseless Age 2 (July 1897): 2–4.

distance between watering stops for a noncondensing steamer), yet charging still took considerably longer than decanting water or gasoline into storage tanks. Also, over time, boosting batteries with repeated quick charges reduced capacity and total service life.[32]

Second, there was no standard battery plug among either vehicle manufacturers or local utilities, so finding an accessible electric plant did not guarantee the ability to instantly "fill 'er up." While urban electric vehicle enthusiasts discussed the development of an electrant—a curbside electric hydrant that would dispense a set amount of electricity for a modest charge (see figure 5.2)—a technical report on the 1900 Madison Square Garden auto show bemoaned the proliferation of plug designs.[33] Outside select urban areas, the prospect of a standard connection was even more remote. Maxim's 1897 report, for instance, did not claim that public charging facilities were actually available in the cities he listed, only that the existence of central stations meant that in principle an electric vehicle driver could recharge there. Actually connecting a vehicle to the local power system could require creative solutions, as when Maxim was forced to run a wire directly from the station busbar through an open window into the vehicle battery compartment.

Third, recharging a vehicle battery required more than finding a generator and compatible wiring. Because early electric vehicles did not operate at a standard voltage, rectifiers and converters often had to be inserted into the charging circuit. Under these conditions, the

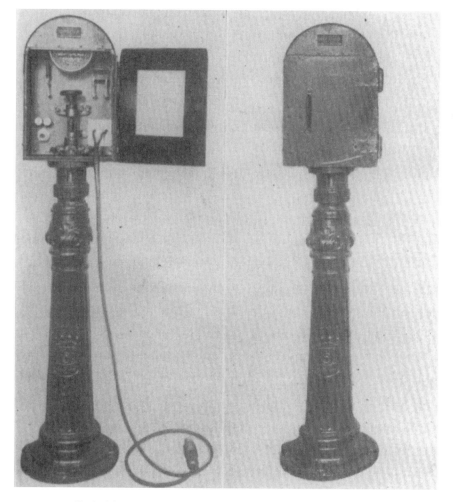

FIGURE 5.2. Curbside electric charging station, 1914. *Source:* Electrical World *65*
(January 16, 1915): 179.

vehicle owner also had to serve as in loco station engineer. Practical
knowledge of electricity—how to calculate a battery charge or repair
a short circuit—was scant at best. Unlike steam and gasoline engines,
the storage battery and electric motor were sealed black boxes that
did not yield their secrets to the casual observer. As one storage battery expert noted in 1899, "the average intelligent citizen knows that
a steam engine is operated by the driving of a piston back and forth
in a cylinder by the pressure of the steam behind it, or that a wind-
mill is caused to operate by the sails or blades being pushed by the
force of the wind which strikes them; but the operations going on
in the storage battery, both in charging and in discharging, are far

more subtle than any operation in mechanics."[34] Reporting on his nearly 2,000–mile journey from Lincoln, Nebraska, to New York City in 1908, electric vehicle manufacturer Oliver P. Fritchle observed that such a trip could not be made in an "ordinary electric car unless it was handled by an expert electrician."[35]

During the crucial early years of the automobile, the infrastructure for generating and distributing electricity could not support widespread use of electric vehicles outside major towns and cities. Only drivers with extensive electrical expertise would be willing to risk touring in an electric vehicle; and even then, careful trip planning was necessary to insure access to electricity.

From 1897 to the mid-1920s, these fundamental problems continued to limit the feasibility of touring in an electric vehicle, as three different geographical snapshots show. First, at the national level, figure 5.3 shows a map of the national electrical system in 1923. Reproduced from *Electrical World,* the map illustrates both the gradual and uneven diffusion of electricity into rural and outlying areas as well as the lack of coordination and interconnection among local electrical systems. By 1923 some regions of the country were already well connected with each other, but the number of towns still operating their own stand-alone generating stations and the large areas still entirely without service underscore the inability of the electrical system to support a universal electric vehicle.

At the state level, figure 5.4 compares the availability of charging facilities with the number of cities and towns in Massachusetts in 1914. Nine out of ten Massachusetts towns were served by a central station; but vehicle charging facilities were available only in one out of six, and fewer than one in ten could provide the high-power charging facilities needed for midday boosting when touring. Given that by 1914 gasoline could be purchased in nearly all of the 333 cities and towns in Massachusetts, the lack of an effective electrical infrastructure clearly put the electric touring vehicle at a distinct disadvantage.

Finally, at the local or regional level, figure 5.5 illustrates the gradual expansion of electric charging stations in and around New York City from 1901 to 1928. Collected from private lists and public maps and pamphlets, each public access charging station is shown plotted at the center of a circle four miles in diameter; within the shaded areas, an electric vehicle operator was no more than two miles from a charging station. In 1901 and 1906 the electric charging

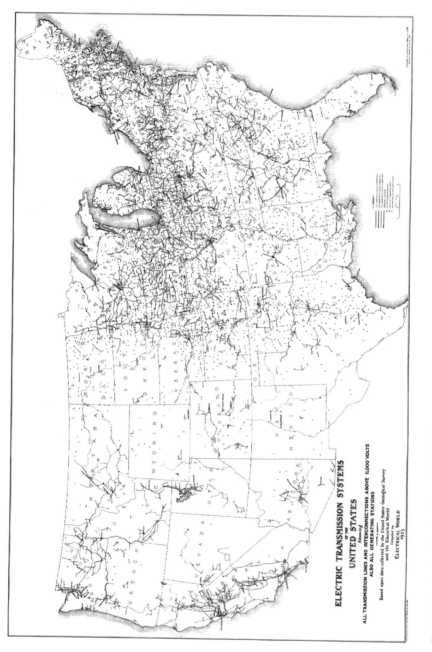

FIGURE 5.3. The national electrical system, 1923. *Source: Insert in Electrical World 83 (January–June 1924), 1.*

infrastructure could not support long trips outside Manhattan. By 1915 many gaps in coverage remained, but the service area had increased considerably. Significantly, however, trunk lines connecting New York City with regional centers such as Albany, New Haven, and Philadelphia were still missing. Relatively few stations were added after 1915; so as of 1928, the situation was not much improved. The absence of competition in the charging station market posed a further complication for the electric vehicle. As *Horseless Age* correspondent Harry Dey wrote in a 1905 article, stations were sufficiently scattered to accommodate the electric vehicle, but "if a customer is dissatisfied at one place, in many cases he will have to go quite a distance to find another." Level of knowledge and quality of service varied widely, prompting Dey to conclude that while some charging stations employed expert electrical engineers, others tried "to get along with electric bell hanger graduates."[36]

In the New York City area, as in the rest of the country, electric vehicle owners faced a paradox. If they desired competitive service for a limited range, urban-based electric vehicle, the local charging infrastructure was too spread out; yet if they wanted minimal service but extended coverage for possible electric touring, the charging stations were not widespread enough, nor did stations support routes between important destinations. The expansion of the electric vehicle charging network during the first two decades of the century demonstrated the growth in the use of electric vehicles, especially before 1914. But the charging network, despite its expansion, was unable to keep pace with either the spread of population or the demand for increased range to tour. Neither fish nor fowl, the electric vehicle charging infrastructure was caught between competing visions of the role of the electric vehicle.

ELECTRIC TOURING

With the existing infrastructure favoring internal combustion vehicles for touring, the question naturally follows, "Why did supporters of the electric vehicle pursue the idea of the electric touring car?" Not surprisingly, members of the electric vehicle community asked themselves the same question. Their answers took several overlapping forms.

First, electric vehicle supporters recognized the cultural appeal of touring. The idea of using the motor vehicle for touring was present almost from the birth of the industry, and touring quickly emerged as the preferred use for the privately owned internal combustion automobile. German scholar Wolfgang Sachs and American intellectual historian Stephen Kern have described the allure of speed, and Dutch historian of technology Gijs Mom notes the unique suitability of the gasoline vehicle for this application. In Mom's view, the internal combustion vehicle was an "adventure machine": like a first-rate stallion, the early gasoline car was challenging to operate and control, and the ability to do so evidenced the skill of the driver. Despite its mechanical lineage, the touring car brought its occupants closer to nature, and the early gasoline vehicle was sufficiently prone to mechanical malfunction yet simple enough to repair that drivers could trust both that the vehicle would break down and that they would be able to demonstrate their mechanical talents by repairing it. As Baker vehicle representative E. J. Bartlett observed in 1914, "there was perhaps a feeling of pride in being able to drive a gasoline car and to fix it if it went wrong."[37] By contrast, the electric vehicle was "the opposite of spectacular"; its relatively quiet performance was "not impressive when contrasted with that of its noisy coworker."[38] The small group of electric vehicle enthusiasts who pursued the dream of electric touring, in effect, sought to create an electric "adventure machine." To them, touring by electric vehicle offered the same benefits as touring by internal combustion; it was fun.

Second, individuals and firms saw electric touring as a way to gain competitive advantage over rival producers of electric vehicles. After completing a long cross-country journey in his electric vehicle, Oliver P. Fritchle, owner of the Denver-based Fritchle Automobile and Battery Company, reported that his "100–mile Fritchle electric" had averaged ninety miles per day traveling through the Allegheny Mountains.[39] Claiming that much of his trip had covered the very same route as had been followed by the 1908 Glidden Tour, Fritchle tried to enter one of his electric vehicles in the 1909 Glidden Tour. These tours, named after telephone executive and automobile enthusiast Charles J. Glidden, showcased the reliability of the automobile for long-distance touring. Speeds were strictly limited, and entrants won points by not requiring service on the road.[40] Along similar lines, Bailey and Company hoped that Bailey's spectacular touring accomplishments would attract customers to the Bailey line of electric vehicles. The actual effect of these events on sales is unknown, but

Figure 5.4

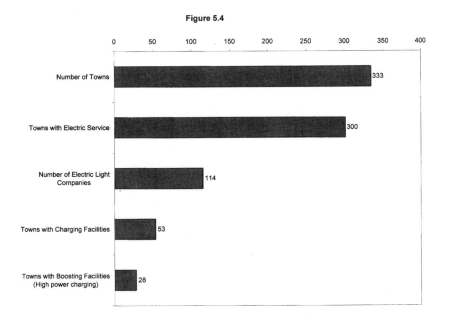

FIGURE 5.4. Availability of charging facilities in Massachusetts, 1914. *Source: J. S. Codman, "Touring by Electric Automobile,"* Electric Vehicles *5 (July 1914): 23–24.*

claims about range certainly played an important part in the marketing and selling of electric passenger vehicles.

Third, electric touring, like racing, validated the capabilities of new technology. To this end, electric vehicle builders and drivers promoted touring as a test of a given vehicle and competed for bragging rights about whose vehicle could go farthest on a single battery charge. In one of the earliest examples of this behavior, Hiram Percy Maxim and fellow engineer Justus B. Entz, who later became known for developing an innovative electromagnetic transmission system, spent several days in the fall of 1899 driving a specially constructed electric runabout back and forth between Philadelphia and Atlantic City. On November 17, in an elaborately staged event using handscribbled notes tossed from the speeding vehicle to prepositioned monitors along the side of the road, the two men covered one hundred miles on a single charge at an average speed of 12.9 miles per hour with a little bit of juice to spare.[41] Four years later, in October 1903, President Charles Edgar of Boston Edison approved a long-distance demonstration of the new Edison alkaline battery. Over

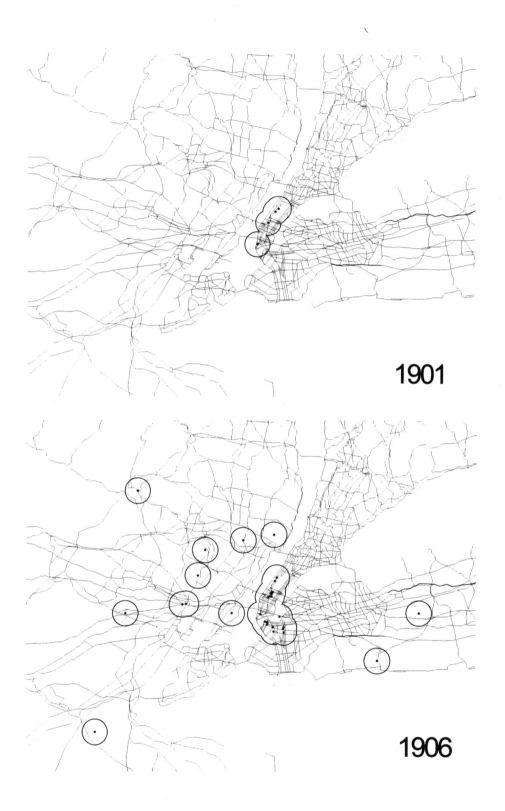

1901

1906

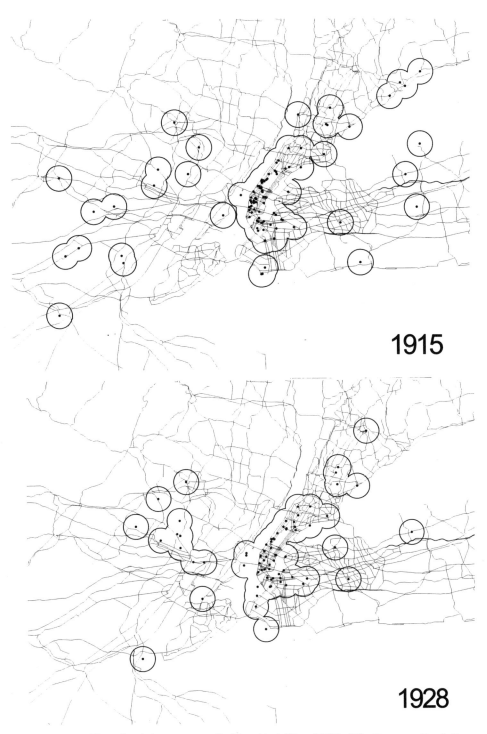

1915

1928

FIGURE 5.5. Charging infrastructure in New York City, 1901–28. *Sources: Frank C. Armstrong Papers, Henry Ford Museum and Greenfield Village, Dearborn, Mich., box 1, folder 1; and the map collection of the New York Public Library, prepared in digital form with the assistance of Shauna Mulvihill.*

four days of "leisurely travel," an Edison employee, H. M. Wilson, drove a stock Columbia electric service wagon 249 miles from Boston to New York City at an average speed of ten miles per hour.[42] Although the battery was charged several times en route, the vehicle managed to travel fifty-three miles on a single charge from Worcester to Springfield. By 1909 Emil Gruenfeldt of the Baker Motor Vehicle Company in Cleveland claimed to have covered 160.8 miles on a single charge, averaging 13.4 miles per hour.[43] Two years later, he beat his earlier mark by traveling 201.6 miles on a charge at the slightly slower average speed of 12.75 miles per hour.[44] No organization kept track of these putative records, so it is difficult to know whose vehicle really was capable of going farthest on a single charge.[45] In practice, it did not matter who actually held the record. The single-charge trips were staged events, run at slow and constant speed over carefully selected routes.

Finally, electric touring built awareness of the electric infrastructure and focused attention on the need for expanding it. Reporting on Wilson's 1903 ride from Boston to New York, *Electrical World* noted the "general ease and facility with which charging arrangements can now be found along this and other routes."[46] Elsewhere, in 1910 F. A. Babcock of the Babcock Electric Carriage Company took a 463–mile, four-day electric tour through western New York State to get "in closer touch with" the power companies in the area, all of whom were "greatly interested" and "promised to equip for the care and charging" of electric cars.[47]

Within the electric vehicle community, touring in general and traveling long distances on a single battery charge in particular were controversial subjects. The electrical fraternity had long argued that speed and range (crucial determinants of touring ability) were irrelevant when it came to general passenger vehicles. For example, reports in the electrical press on the 1895 Paris-Bordeaux-Paris road race noted that while electric vehicles had not fared well in the race, it was "generally conceded that the electric vehicle in urban service, or for use where the mileage limitations of storage batteries need not be considered, has no rival."[48] Later, a 1901 *Electrical World* editorial, reporting on an endurance run sponsored by the Automobile Club of America, criticized the organizers and the participants for letting the event degenerate into a "scorching match" where endurance "was mostly in the hands of the express companies in the form of spare parts." On the issue of speed, the editorial was equally blunt:

"a motor car with a possible speed of a mile a minute [i.e., sixty miles per hour] is a striking engineering feat, but it is not a business, and its field of usefulness is a small and inconsequential part of the region that must eventually be tenanted by motor vehicles." The journal was intimating that surely no one would need to travel sixty miles an hour in an automobile; therefore, do not expect a sustainable market for such a novelty. Rather, the automobile was "destined to replace the truck and the cab and the trap rather than the sulky."[49]

Other leading voices also weighed in on these issues. Electric vehicle pioneer Hayden Eames noted that "century trips" (single-charge trips of one hundred miles or more) were not to be expected from "stock electric vehicles" and criticized the "questionable practices" that were producing these long-distance runs. In most instances, he claimed, long runs were accomplished using special tires that "would be utterly unserviceable in regular duty" under conditions that "are made to order for long distance travel." For Eames, these performances were merely intended to "deceive the unwary"; and as a result, many "credulous people" suffered from the misguided belief that electric vehicles "should be capable of traveling 100 miles on a single battery charge."[50] In 1916, as the electric vehicle renaissance was faltering in the face of the Great War, *Electric Vehicles* weighed in on these same issues with an editorial titled "The Unimportance of Touring." Citing as "proven fact" the claim that 99 percent of the usefulness of a vehicle "lies within a radius of less than thirty-five miles," the journal claimed that buyers needed to be educated with "facts about long distance touring."[51] Some in the electric vehicle industry tried: one cooperative advertisement by electric vehicle dealers in Columbus, Ohio, showed a map of the region with concentric circles at fifteen- and thirty-mile radii (see figure 5.6) and asked, "How often have you motored beyond this territory?"[52] Unfortunately, such appeals to the mundane reality of everyday life did not sell cars. Customers did not want to hear that touring was more difficult and strenuous than they imagined, that "touring is a delusion," that "the ability to do 70 miles an hour is of no possible use to them," and that "eighty percent of the work done by gasoline passenger cars could be better done by electrics, and more economically."[53]

Even if all the claims by electric vehicle supporters were true, and typical drivers would never or rarely take advantage of the touring ability of their gasoline car, consumers still invested in the "dream of long distances." *Electric Vehicles* accurately diagnosed the problem:

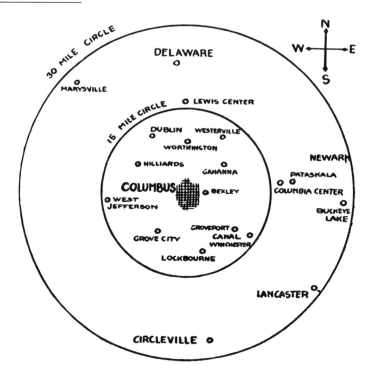

How often have you motored beyond this territory?

FIGURE 5.6. Cooperative advertising on the unimportance of range for typical automobile drivers, Columbus, Ohio, 1916. *Source:* Electric Vehicles *9 (August 1916): 50. Reprinted by permission of the Linda Hall Library, Kansas City, Mo.*

"the things a gas car can do, but seldom does, are the things a brand new driver thinks he wants to do and the things an electric will not do. By the time he finds out he was mistaken in his desires, he already owns a gas car and is apt to continue along the line he has started."[54] Regardless of the actual importance of speed and range, their perceived importance encouraged manufacturers of electric vehicles to continually strive to improve their vehicles' performance so as to compete with the gasoline passenger car. Moreover, as we learned in chapter 4 with the emergence of the universal commercial vehicle, a driver might still choose to purchase a gasoline vehicle for that 20 percent or 10 percent or even 5 percent of trips that the electric vehicle could not manage.

The paradox of electric touring was that while the electric vehicle industry was claiming that touring was unimportant, dozens of industry leaders were spending inordinate amounts of time, energy, and money arranging and participating in the very spectacles that other industry leaders were inveighing against. For each achievement, dissenting voices claimed that the achievement was in fact a distraction.

The electric vehicle was not a touring vehicle because you were always "slave to a wire." One could make the wire longer by extending the range somewhat, establishing remote charging facilities and even creating mobile charging stations, but that still did not turn the electric vehicle into an "adventure machine." The spirit of touring was inconsistent with the need to drive strategically, to always be thinking about available range and the location of the nearest charging facility. So why did some electric vehicle enthusiasts persist in trying to position the electric vehicle as a touring vehicle when their own industry leaders were telling them that it was a waste of time and effort? Electric vehicle drivers made a virtue out of necessity by turning the practice of strategic driving, of having to plan exactly how far one could go with a limited amount of stored energy, into a challenge unto itself. Electric vehicle drivers channeled the "racing spirit"— described in the pages of *Electrical World* as always seeking a way to compete, even if only by racing wheelbarrows—in new directions by competing over how many miles a vehicle could travel on a single charge of electricity, how fast it could go, and how far it could tour. Whether Colonel Bailey seriously expected to establish the practicality of cross-country electric touring is unclear. But regardless of the stated motives of Bailey and his cohort of fellow travelers (expanding the performance envelope of the electric vehicle, building awareness of the electric and road infrastructure, and increasing brand awareness for their respective companies), electric touring was primarily a personal challenge, not unlike touring by internal combustion.

THE DYNAMICS OF AUTOMOTIVE SYSTEMS CHOICE

As *Horseless Age* columnist Albert Clough recognized in 1902, choosing an automobile required that prospective drivers take account of more than mere technical characteristics. The automobile was part of the larger transportation system and depended on available infrastructure. Since electrification was still in its infancy, one could hardly afford to be "slave to a wire."

Unlike traditional explanations of the process of technological choice, this chapter has shown that if touring was the desiderata for the automobile, then internal combustion—a technology that could take advantage of the prevailing infrastructure and existing modes of energy distribution—was the preferred solution. Just as the typical sport utility vehicle of the mid-1990s is rarely driven off road but is purchased for the lifestyle image it embodies, early motorists valued the image of touring more than its reality. And some manufacturers such as Bailey and Babcock even produced electric roadsters that captured the image of the touring vehicle.[55]

Supporters of the electric vehicle responded to the touring impulse in one of two contradictory ways. Some, like Maxim, Fritchle, and Bailey, set out to prove that electric vehicles were capable of touring. Others (Eames, editorialists, and most of the other vehicle manufacturers) ignored the touring market and targeted owners of traditional horse-drawn transportation. These parallel efforts created confusion about the appropriate role of the electric vehicle. Were potential electric vehicle customers buying a long-distance touring vehicle? If not, then why were some manufacturers touting it as such?

As the data on infrastructure development show, this confusion extended into the physical realm of roads and charging stations. Was the electric an urban passenger service vehicle, as the owners of the Electric Vehicle Company intended? Or was it an affordable commercial vehicle, as the members of the Electric Vehicle Association of America and the light and power companies believed? Or was it destined to ultimately become the universal vehicle, as the behavior of Bailey, Fritchle, and other true believers suggests? During the first two decades of the automobile age, the infrastructures developed to support the electric vehicle reflected this confusion. For most motorists, electric touring was putting electrons before the wire. No doubt Colonel Bailey enjoyed his limerick and his potted plant, but they were no substitute for a fully functioning network of charging stations to support electric touring.

EXCURSUS: THE ACCIDENTS THAT NEVER HAPPENED

> All the storage battery experts and the heads of the leading
> storage battery companies of this country at that time [1897]
> did not believe that storage batteries were applicable to auto-
> mobiles. They did not know anything about their application to
> automobiles. The only people who ever tried to apply them to
> automobiles on any scale were Messrs. Morris and Salome
> [*sic*], and the very first battery used showed that the situation
> was not understood, although they perhaps understood it as
> well as anybody else. *The things we were looking for as acci-*
> *dents in the automobile have never happened.* Any quantity of
> other things have happened; any quantity of them, but those
> particular things did not happen.
>
> Hayden Eames, April 1907[1]

In part 2 a picture of the electric vehicle has begun to emerge. Companies, organizations, and systems dedicated to the electric vehicle
were conceived and launched amid great fanfare and high hopes: the
EVC was to provide passenger service in cities across the country and
beyond; the EVAA would unite the disparate participants in the electric vehicle system behind a shared vision of urban electric service.
All presumed that the model of separate spheres, in conjunction with
sound marketing and common sense, would produce an efficient
mixed system to meet commercial needs. Despite the misgivings of
some electric vehicle supporters, even the struggles surrounding electric touring seemed destined to foster the growth of a dense network
of electric charging facilities that would ultimately allow modest electric touring. At each juncture, electric vehicle supporters believed that
they were poised to succeed, that the future would unfold according
to their beliefs and expectations. Yet each time, in each domain, the
efforts resulted in failure.

At every point, we have also entertained a series of counterfactual
hypotheses about what might have happened had the ventures succeeded. Clearly, this mode of inquiry is problematic. Discovering what
actually happened is challenging enough, so why confuse matters by
looking at what did not happen? This counterfactual style of argument will doubtless strike some readers as needlessly speculative. In
defense of the technique, however, part of its purpose is to recapture

the full range of alternative outcomes and, by so doing, isolate those that were patently ridiculous from those that were unlikely but not unthinkable. Against the background of ninety years of automotive history, one needs a measure of indulgence to even entertain the idea of an alternative to internal combustion. To imagine an electric vehicle–based transportation system, we have no choice but to speculate.

Moreover, as theorist Jon Elster has argued, historians constantly engage in counterfactual reasoning without knowing it. Whenever we ascribe causality to one set of factors over another, we create an implicit counterfactual claim about other possible outcomes. But on a deeper level, the logic for engaging the counterfactual domain begins with potential turning points—"natural joints," as Elster calls them. Rather than fabricate alternate scenarios out of whole cloth, our hypotheses have been rooted in outcomes that participants themselves believed were reasonable, if not likely, to occur based on their expectations about the future. In effect, these natural joints minimize the need for ex post theorizing and help establish legitimacy for the counterfactual scenarios.[2]

In this instance, examining what earlier generations thought would happen is more than mere intellectual hand wringing. Expected outcomes—counterfactual alternatives—shaped real behavior. As Hayden Eames's quotation clearly demonstrates, what did *not* happen—the "accidents" that were looked for—is part of the story, not simply the concluding footnote to a host of unfulfilled dreams and failed ventures. From this perspective, his statement takes on new meanings. Had we not explored a wide range of natural joints in the evolution of the automobile, the temptation would be to think of the accidents that did not happen in simple, discrete terms: if a radical innovation in battery technology had occurred, the electric vehicle would have fared better and, depending on when exactly this innovation took place, might have been able to establish a dominant market position in important sectors of the expanding automobile industry (perhaps for urban passenger service, commercial delivery, or touring). Because we have developed a broader view of possible alternate scenarios, however, we see that Eames and his original cohort of electric vehicle pioneers anticipated any number of possible accidents with complex links to each other and to the development of the automobile industry. From battery-powered electric cabs and buses, to durable separate spheres, to highly developed systems for battery exchange, the first generation of electric vehicle pioneers foresaw a range of possible electric pathways. Yes, they may also have expected a radical breakthrough in battery technology. But the preceding chapters have expanded our understanding of how the participants perceived the early history of the industry, alerting us to the full range of possibilities.

The chapters that follow explore the potential implications of some of this history for our understanding of present and future paths of technological change. Although historical events and issues are present throughout, interpretation and the search for perspective will now come to the fore. What accidents are we waiting for today, and why?

PART THREE

PERSPECTIVES

6

THE BURDEN OF HISTORY
Expectations Past and Imperfect

> One trouble with the electric vehicle has been that too much
> was expected of it in the early stages of the art. Various com-
> panies rushed into its manufacture, but, strange to say, stor-
> age batteries did not alter any of their previously well-known
> characteristics and objected to neglect and mistreatment just
> as much when placed on an electric vehicle as when used in
> any other way. Almost as much improvement has been made
> in batteries and vehicles as has been made in gasoline auto-
> mobiles, but many of the old traditions still seem to cling to
> the electric vehicle. One of these is the fallacy that a large
> mileage on one charge is essential to ordinary business about
> town. The average electric vehicle, as built today, has consid-
> erably more available mileage on one charge of battery than
> the average vehicle of 10 years ago, and what is more, has a
> considerably greater mileage than is actually needed in the or-
> dinary run of business or pleasure, except where a long tour
> is undertaken.
>
> Editorial, *Electrical World*, March 1909[1]

By 1909 the leading editorial voice for the electrical industry was already well versed in the art of apologizing for the liabilities of the
electric vehicle. From the first, apparently, "too much was expected
of it." Despite significant progress in many of the technologies that
made up the electric vehicle system, none of the electric vehicles—
passenger, commercial, or touring—were able to deliver on expectations.[2] By contrast, during this period the internal combustion vehicle
system witnessed unprecedented improvements that exceeded every
possible expectation.

What role did expectations play in the development of the electric vehicle? Indirect evidence suggests that high initial expectations
may have actually slowed its spread, leading, in turn, to lower rates
of innovation within the industry. The way in which the evolution
of the turn-of-the-century electric vehicle was shaped by expectations
about technological change also provides a crucial link between past
and present. Echoes of the historical dynamic of expectations and
behavior that contributed to the collapse of the early electric vehicle

can be heard in recent debates about the future of the electric vehicle. In this sense, from the 1960s through the mid-1990s, the
electric vehicle has continued to bear a burden of expectation that
explains, at least in part, the challenges and complications surrounding attempts to reintroduce the technology into the turn-of-the-
millennium transportation system.

IF NECESSITY IS THE MOTHER OF INVENTION, WHERE IS THE BETTER BATTERY?

Early electric vehicle enthusiasts had many reasons to hope for a revolutionary breakthrough in energy storage technology; their generation had lived through an age of technological miracles. Engineers
working in the last quarter of the nineteenth century had witnessed
multiple technological revolutions, from the telephone to the automobile to electricity itself. As late as the Great Depression, working
managers in the electrical industry could personally remember a time
before the commercialization of electricity. A generation after Thomas Edison first sold electricity from his Pearl Street station, many
were entitled to believe that a better battery was indeed "only a day
away." Initially, faith in the imminent solution to the battery problem ran high. Over time, however, hope gave way to steadfast optimism or wistful resignation; expectations were never fulfilled, even
though incremental technological changes dramatically improved the
capabilities of the typical electric vehicle.

   In an important contribution to the study of technological
change, economic historian Nathan Rosenberg has observed that expectations play an important role in shaping the rate and direction
of inventive activity and the spread of new technology. He concludes
that under certain conditions fear of technological obsolescence can
slow the advance and spread of new technology, even and especially
during periods of rapid technological advance.[3] As Harold Passer
notes in his analysis of the early electrical industry, "the manufacturer has to convince the prospective buyer that no major improvements are in the offing. At the same time, the manufacturer must
continue to improve his product to maintain his competitive position and to force existing products into obsolescence."[4] Any contemporary high-tech consumer is acutely aware of this phenomenon. The
entire high-technology product life cycle (described by diverse writers such as William Abernathy, Eric Von Hippel, and Geoffrey Moore)

is predicated on rapid technological obsolescence.[5] Early adopters—those among us who for whatever reasons simply must have the latest, lightest, "best" version of the newest gizmo—buy first, regardless of price. But the ultimate success and profitability of a given product depends on whether or not that product is able to, in Moore's phrase, "cross the chasm" to reach mainstream consumers. For instance, the market for the first digital wristwatch, the Pulsar Time Computer offered in 1972 by the Hamilton Watch Company for the exorbitant price of $2,100, was limited to those few lead adopters who simply had to have it. Many consumers were willing to wait a few years for the price to drop by a factor of ten, and more still were willing to wait a few more years for the price to drop by a factor of 100. By the end of the decade, the product life cycle had run its course; not one U.S. firm was even manufacturing digital wristwatches, Hamilton had failed, and *Time* magazine was giving away watches as subscription premiums.[6]

In the case of the early history of the electric vehicle, expectations shaped both technology and the market for it. Although by 1902 the lead-acid battery was an adequate if unwieldy source of electric power, many industry experts expected radical improvements in storage battery technology. From the 1890s on, there was a persistent demand—or "need"—for a significantly lighter, more powerful, and more durable alternative to the lead-acid cell. Amid the wondrous changes of the late nineteenth century, surely it was only a matter of time before a Bell or a Westinghouse or an Edison or some other innovator would develop a vastly improved battery. Edison himself aspired to claim the prize with his alkaline nickel-iron battery design.[7] But even before Edison announced (prematurely, it turned out) his "miracle" battery, electrical industry experts were anticipating dramatic advances in electrical energy storage. For instance, in 1897 *Electrical World*, in its first editorial devoted to the horseless carriage, praised Morris and Salom's "electric hansoms" and noted their "great and immediate popularity" yet also predicted rapid advances in battery technology: "There is not the least reason . . . to assume that the accumulator [battery] will not undergo great and marked changes and improvements. In general, whenever there is a demand for a specific improvement or invention, it soon makes its appearance. There has already been a great increase in output and capacity in accumulators, and it is certain that much room for improvement remains."[8] The next month, another editorial echoed this theme: "The success of electric automobiles will unquestionably cause a desire on the part of their users for a longer radius of travel, which only means an accumulator of greater capacity for a given size and weight. This is almost certain to be produced when the demand for it is sufficiently urgent."[9] During this initial phase, expectations were expressed as simply that: expectations.

Two years later, in 1899, with a replacement for the lead-acid battery still nowhere in sight, the electrical journals ratcheted up the stakes, noting the need for a better battery. Summarizing the state of the "electromobile," *Electrical Review* concluded: "It has been said that no great invention waits long to appear when the demand for it becomes sufficiently strong. If this is true, it will not be long before a lighter accumulator is produced. There is hardly any improvement in electrical apparatus so greatly needed. Given a light, mechanically strong accumulator that will stand heavy charging and discharging, the electromobile problem becomes comparatively simple. . . . Here is a problem that ought to interest all inventors and experimenters. The market for such an accumulator is ready-made and will be enormous."[10] By 1900 *Electrical Review* was framing its expectations more directly: "It has been claimed over and over again that whenever the conditions absolutely require a new invention the invention is forthcoming. Never before in the history of the electric arts has there been a more insistent demand for anything than there is now for a storage battery of higher capacity per unit of weight; to put it in other words, the light storage battery is urgently needed."[11] Finally, in 1901 the journal articulated its true view of invention: "the demand for a proper automobile storage battery is so crying that it soon must result in the appearance of the desired accumulator. Everywhere in the history of industrial progress, invention has followed close in the wake of necessity."[12] These statements underscore the extent to which leading voices within the industry anticipated greater improvements in battery performance than inventors could actually deliver. Although these same journals would later recant their insistence on the need to replace the lead-acid battery and affirm its adequacy for most vehicle applications, the early rhetoric may well have depressed consumer demand for electric vehicles because prospective customers postponed buying electric vehicles in the belief that a radically better battery solution would soon arrive.

Electric vehicle supporters were continually forced to defend, justify, and explain away the perceived liabilities and unfulfilled promises of the first generation of vehicles. Numerous accounts attributed the "indifference" to the electric vehicle to "the unfortunate experience of ten years ago [1900], when a somewhat sudden movement towards its general introduction had an unpleasant result."[13] In Boston the early experience "with imperfect and undeveloped apparatus . . . [had] resulted in a large amount of conservatism in the local community."[14] In Brooklyn an observer noted that "visionary claims of early manufacturers" and "their unfulfilled promises" militated against the potential introduction of new electric vehicles and added that he knew of no fewer than eighty-three "inoperative old electric vehicles whose owners, owing to past experience, would not entertain propositions for the modern electric vehicle at any cost."[15] "The electric vehicle," *Electrical World* observed in its annual overview of the electric vehicle industry in 1910, was "distrusted owing perhaps to discouraging experiences by owners upon its advent some years ago."[16] Apparently, the legacy of the Lead Cab Trust had not faded with the passage of time.

In fact, as the motor vehicle continued to evolve, this legacy of unfulfilled expectations emerged as the cornerstone of the popular history of the electric vehicle. In 1917, for instance, an *Electrical World* editorial titled "Thinking Straight on the Electric Vehicle" explored both the unmet expectations and the surprising "tenacity" of the electric vehicle: "The electric started off on the wrong foot as a competitor of the gas car for all purposes. As a result, it is still walking through life with blackened eyes."[17] The same theme was present in a 1924 article based on a survey of twenty-five leading central stations. After promoting the "splendid" ability of the electric vehicle to provide local urban transportation service, the report attributed the "apathetic response to suggestions that electric vehicles are excellent transportation agencies" to the experience of the early days: "the electric battery vehicle had a premature development. . . . The result was dissatisfaction and failure and the promulgation of false conceptions which persist to the detriment of the electric vehicle even today."[18] The public both remembered the failures and raised its sights for future vehicle generations. Twenty years after the Electric Vehicle Company and its competitors first failed to meet expectations, the legacy of the first generation of electric vehicles continued to cast a long shadow over the industry.

How many electric vehicles were *not* sold because potential buyers were anticipating improvements in technology? It is difficult to measure the impact of expectations on consumer behavior, but two types of records suggest that expectations did play an important role in the development of the electric vehicle market. First, the history of false starts surrounding the introduction of the Edison alkaline storage battery heightened expectations and encouraged potential electric vehicles customers to defer buying vehicles. Initially announced by Edison in 1901 as the long-sought alternative to the lead-acid battery cell, the first Edison type E cells were not available until 1903. Little over a year later, however, these batteries were recalled. Not until 1909 was the redesigned type A battery cell available on the market.[19] During this period, many would-be electric drivers either bought no car at all or bought an internal combustion vehicle. As one participant at the March 1909 meeting of the Pacific Coast Electric Automobile Association observed, "the unwarranted promise by the daily newspapers of a 200–mile battery has proved a serious obstacle to the introduction of electric vehicles."[20] Edison, the "wizard of Menlo Park" and a master promoter of his own achievements, had whetted the public appetite for a miracle battery but was unable to deliver on his promise.

Second, responses to a 1914 report that Henry Ford and Thomas Edison were joining forces to market a reliable, low-priced electric car provide an indirect measure of consumer expectations. Announced at the New York Automobile Show in January 1914, the Ford-Edison vehicle promised to have a range of one hundred miles and cost between $600 and $1,000, well below the average price of the typical electric vehicle of 1914.[21] A thick sheaf of letters in the archives of the Ford Motor Company testifies to the demand for such a vehicle. From across the country, customers, dealers, and suppliers wrote expressing support for the venture and seeking details about the car. Again, consumers were eager for an electric vehicle that could deliver on past promises; but Ford, recognizing the difficulties that lay ahead, chose instead to work with Edison on developing an electrical system for the Model T, which did not feature an electric starter until 1919.[22]

Early lead-acid batteries, although adequate for stationary applications, were admittedly not well adapted for traction uses. Over time, innovations such as the Exide and Exide Ironclad designs—developed and introduced in response to market conditions (that is, "need")—resulted in considerable cost and performance improvements. Similarly, the Edison "miracle" battery represented an important advance over first-generation vehicle batteries; it was lighter, more durable, and better suited to boosting. But none of these advances delivered on the ultimate promise of technological transformation; they all failed to meet expectations. Although battery technology continued evolving to the point where industry observers could claim victory— as when an *Electrical Review* columnist observed, "The much maligned battery will stand its share if it is only given a fair chance"—the technology never succeeded in shaking free of its early history of failed promise.[23]

MOVING TARGETS: THE CHANGING CAPABILITIES OF ELECTRIC AND INTERNAL COMBUSTION VEHICLES

One of the key implications of adopting a path dependent approach to the study of technological evolution is that relatively small differences in initial performance between electric and internal combustion vehicles were magnified by subsequent events. Thus, the differences in performance between typical electric and gasoline vehicles in 1900 were relatively modest, but by 1914 the contrast was more marked. The electric vehicle of 1914 was no longer competing against a crude, unreliable, gasoline-powered horseless carriage. Rather, by 1910 the internal combustion vehicle industry had itself evolved. Leading firms such as Ford, Buick, and Studebaker were producing many thousands of vehicles each year. Numerous advances in design, technology, and manufacturing had propelled the industry forward. As of 1914, therefore, the electric vehicle industry confronted the following dilemma: the electric vehicle of 1902 (that is, after the initial kinks had been worked out of the Exide battery) was actually more acceptable to consumers than was the electric vehicle of 1910. In absolute terms the electric vehicle of 1910 was vastly superior to the first-generation vehicles produced at the turn of the century; but relative to both expectations and the internal combustion vehicle of 1910, the passenger electric car was actually further from commercial viability than was its predecessor.

Chapter 5 reviewed some of the stunts intended to demonstrate the touring capabilities of electric vehicles. For all practical purposes, however, the electric vehicle lost ground to internal combustion during the first fifteen years of the industry's life. In 1900, for instance, *Electrical World* concluded that the electric vehicle was acceptable for touring: "for pleasure riding a machine that can make fifteen miles an hour if necessary or run forty or fifty miles without attention and with the reasonable certainty of getting home is far more useful than a racer which may make forty miles in an hour, or be left on the road and sent home by freight."[24] Setting aside the infrastructure issues we have already examined, we see that a typical electric vehicle of 1900 could almost meet these requirements. For example, EVC fleet cabs averaged nearly thirty miles per vehicle and were hardly optimized for long-range travel. Ten years later, many aspects of electric vehicle performance had improved. Even *Horseless Age* defended the capabilities of the electric vehicle of 1910: service range had increased to eighty miles "at speeds approaching twenty miles per hour"; and other advances in design, tires, and storage batteries led the author to conclude that "many of these improvements consist only in a change in dimension, and seem of no importance when described in print, yet their aggregate result has been to make the electric runabout a thoroughly practical vehicle for business and pleasure driving in urban and suburban districts."[25]

But by 1910 the ground rules had changed. An article in *Central Station* on the new Bailey long-distance electric automobile pleaded, "Give us an electric passenger automobile that will travel 100 to 150 miles at a reasonable speed on a single charge of the battery, do 100 miles at 20 miles per hour, and be able to run in sprints at 30 miles per hour, and hundreds will change from gasoline to the electric vehicle."[26] *Central Station* claimed that the new Bailey roadster could meet these requirements. But regardless of the actual performance of the Bailey vehicle, the minimum conditions for a successful touring vehicle had risen during the decade. Both maximum speed and minimum range had increased, raising the hurdle to the creation of a successful electric touring vehicle.

The question of reliability should have favored the electric vehicle, and considerable anecdotal evidence indicates that the electric vehicle was indeed a dependable piece of machinery. In 1901 Charles J. Glidden recorded 4,000 miles on his electric vehicle, all within a fifteen-mile radius of Boston; only twice was he forced to abandon his vehicle "on account of failure of the power" (in other words, 0.5 failures per 1,000 miles).[27] Commercial fleet data from the EVC claimed that reliability had improved from 6 tows per thousand miles in 1897 to 1.31 tows per thousand miles in 1902.[28] By 1912 truck failures had reportedly dropped to 0.12 per thousand miles, while the same article claimed that gasoline trucks required service nearly three times as often (0.33 breakdowns per 1,000 miles).[29] Thus, in real terms the electric vehicle experienced a fiftyfold improvement in the rate of battery failure between 1897 and 1912, but over the same period of time internal combustion had nearly matched the electric on its own turf. Whereas few would have even dared to use an internal combustion vehicle for commercial service in 1897, by 1912 the reliability of internal combustion trucks was fast approaching that of the electric vehicle.

Not only were early electric vehicles incapable of meeting expectations, but the success of internal combustion created a moving target. As the internal combustion bandwagon gathered momentum, the threshold for minimum required performance continued to ratchet upward, thereby solidifying public perception of the electric vehicle as a technological failure.

BACK TO THE FUTURE: HIGH HOPES AND THE BETTER BATTERY BUGABOO, 1967–1997

The role of expectations in shaping the early history of the electric vehicle provides a lens through which to interpret recent efforts to resuscitate the electric vehicle and reintroduce it into the mature, highly adapted, internal combustion–based transportation system of postwar American society. From the mid-1960s on, electric vehicles faced a different set of problems from those they had encountered at the turn of the century. The social context for technological decision making had changed: internal combustion had become the standard to be bested rather than merely another aspirant to the throne; the petroleum industrial complex had grown so large and pervasive that it had evolved its own well-defined set of political and economic costs; and in cultural terms, the symbolic meaning of what it meant to be an alternative to internal combustion had changed. Where horse pace was once a barrier to the reorganization of transportation and work, gas pace has now become thoroughly embedded in the fabric of daily life.

Comparing recent debates about the reintroduction of electric vehicles against the turn-of-the-century legacy of technological choice reveals the persistence of many of the same behaviors toward technological change that characterized the earlier period. Beginning in the 1960s, as concerns about the safety and environmental impacts of the internal combustion automobile first surfaced at the national

level, and later in the 1970s and 1980s, as oil crises and then economic development pushed alternative fuel vehicles onto the national
agenda, the electric vehicle again emerged as a possible competitor
to internal combustion. In 1967, following extensive public hearings
about the potential market for the electric vehicle, a high-level government committee on the automobile and air pollution noted that
while prevailing electric vehicle technology did not permit the development of anything but a limited-range vehicle, "significant technical advances may be expected in the development of improved
electric energy storage and conversion devices."[30] Like their turn-of-
the-century counterparts, these automotive experts would not endorse
the storage battery of the present but were willing to tout the possibilities for the battery of the future. Within a few years, a number of
domestic prototype vehicles were displayed at car shows and other
public exhibitions, but none of the major manufacturers were willing to enter production until the electric car was "fully competitive
with the piston-engine car."[31] Again, expectations framed decisions
about technology; only a dramatic improvement in battery performance would level the playing field for the electric vehicle.

The oil crisis of 1973–74 added another argument in support of
electric and alternative fuel vehicles: energy security. In the years that
followed, several small companies began to produce and market
purpose-built electric vehicles for the passenger automobile market,
and in 1976 the U.S. Congress passed legislation supporting research
and development on electric and hybrid vehicles.[32] Although the
1976 law also called for an extensive (7,500 vehicle) demonstration
program, government and industry experts opposed the demonstration component on various grounds. One Environmental Protection
Agency (EPA) representative testified that "'premature' demonstration
of such vehicles may lead to a negative public reception. . . . if the
currently available vehicles fall short of public expectations for urban and rural use, the long-run potential of this mode of transportation may never be realized."[33] The Department of Transportation also
argued against the planned demonstration of electric vehicles: "If electric vehicles are ever to be competitive with heat-engine powered cars,
it will be necessary to develop high performance batteries."[34]

On the latter point all the experts were in agreement. Development of a superior battery would have to precede the introduction
of electric passenger vehicles: "Until the improvements in battery
technology occur, it would be premature to support a large and costly

demonstration program."[35] The debates about the electric vehicle addressed other concerns (including safety, consumer acceptability, and lack of charging infrastructure), but the greatest stumbling block remained the battery. Many innovative solutions, from new electrochemical couples to mechanical (flywheel) energy storage, were proposed. As the lead editorial of a special issue of *Machine Design* noted, everyone involved in the electric vehicle revival believed that a better battery was just around the corner: "the consensus among EV proponents and major battery manufacturers is that a high-energy, high power–density battery—a true breakthrough in electrochemistry— could be accomplished in five years."[36]

With expectations once again motivating behavior, the battery was placed at the center of the federal research effort. The 1976 law called for the federal government "to promote electric vehicle technologies and to demonstrate the commercial feasibility of electric vehicles." The overall program developed by the Department of Energy, consistent with this goal, called for "effective and orderly stimulation of the market" and for conducting demonstrations in "sheltered environments in which requisite support services will be established . . . and in which the early demonstration vehicles will be carefully matched to suitable missions and uses."[37] But in practice the program was not focused on system building. Funds went to support a series of discrete technological projects, including a prototype electric test vehicle, advanced "test-tube" batteries, and other novel subsystems.[38] Rather than considering the electric vehicle as part of the automotive transportation system and not necessarily a direct competitor of the gasoline car, the 1976 act sponsored a series of potentially valuable drop-in innovations that were intended to allow future electric vehicles to compete directly with gasoline-powered cars.

Yet no electric car since 1902, regardless of battery or drive train, had been able to compete effectively against its contemporary internal combustion counterpart. The automotive industry stood by its claim that the research program failed because of "the immature state of several key technologies that led to economic, performance, and reliability handicaps that were simply too great to overcome."[39] But given that the internal combustion engine had a sixty-year head start, the federal program was doomed to fail. With the 1976 Electric and Hybrid Vehicle Act, Congress wagered that the federal government could achieve in five years what the entire global battery industry had been unable to accomplish since Thomas Edison promised to revolutionize the electric car battery in 1901. It was a bet they were bound to lose. Excessive reliance on basic research intended to produce a radical breakthrough in battery technology prevented the Department of Energy from implementing even modest infrastructure and vehicle demonstration programs. In an uncanny parallel to the events that had shaped the initial competition between electric and internal combustion vehicles, optimistic technological forecasts that the electric vehicle would soon be able to compete with the gasoline car ensured that practical electric vehicles would remain always just around the corner.

The most recent cycle of interest in electric vehicles began in September 1990, when the California Air Resources Board (CARB) enacted regulations calling on the seven largest car makers active in the state to begin selling zero emissions vehicles (ZEVs) starting in 1998. Initially, 2 percent of new vehicles sold were to be ZEVs, rising to 10 percent by 2003.[40] The economic recession of the early 1990s added a third motive for those hoping to develop electric passenger vehicles: economic development. Several nonprofit groups in California, including L.A. Initiative, Rebuild L.A., Project California, and CALSTART, attempted to link the creation of a local electric vehicle industry to the process of defense conversion.[41] Many large government contractors joined CALSTART, a state-sponsored consortium; and their Showcase Electric Vehicle Program produced a one-of-a-kind demonstration car that was displayed at major auto shows around the world.

Nevertheless, nearly a decade after the Air Resources Board issued its mandate, America's car-intensive culture is still debating the future of the internal combustion automobile and of state and federal efforts to encourage alternatives, especially electric vehicles. In a development reminiscent of the Lead Cab Trust, of the EVAA debates in the 1910s, and of the congressional hearings in the 1970s, the secondary storage battery once again became a lightning rod for criticism from industry and from critics opposed to government intervention in the process of technological development.[42] A range of articles took aim at the supposed benefits of ZEVs, claiming that they were really nothing more than "emissions shifting vehicles" and questioning the environmental consequences of using lead-based batteries in the proposed ZEV fleets.[43] Aided by a multimillion-dollar, public relations campaign, opponents convinced CARB to delay the mandate. On March 28, 1996, the board accepted car makers' claims that they needed more time to develop the improved batteries that would ultimately make electric vehicles competitive with gasoline-powered vehicles and agreed to postpone implementation of the initial 2 percent–level mandate. In an odd twist, several weeks after the CARB decision was approved, General Motors announced that its Saturn division would offer the EV1 for lease in selected southwestern markets beginning in the fall of 1996, one model year ahead of the old ZEV mandate.[44]

By adopting a wait-for-the-battery strategy, California state regulators fell prey to what Michael Schiffer has called the "better battery bugaboo."[45] Since the days of the horseless carriage, proponents of electric cars had claimed to be on the verge of producing a superbattery that would render the lead-acid cell obsolete once and for all. Blaming the battery was a time-honored stratagem employed by both supporters and detractors of electric vehicles. Ironically, as we have seen, it has often been the supporters of innovative battery technologies who have done the most damage to the prospects for real change in the domestic automobile market. In the late 1890s none other than Thomas Edison criticized the lead-acid cells then in use for stationary and mobile applications and promised to revolutionize the industry with his "miracle" battery. And in the 1970s it was an EPA representative who opposed a demonstration program that would have put "premature" technology on the road.

The better battery bugaboo notwithstanding, several factors point to major changes in the storage battery industry. Driven by rising demand from the portable computer and consumer electronics sectors and by continuing advances in affordable, integrated power electronics, the industry may finally prove capable of producing batteries that outperform the best lead-acid designs. Several different designs are poised to meet the U.S. Advanced Battery Consortium midterm goals, defined by the automobile industry as the minimum threshold performance standards necessary for producing a marketable electric vehicle.[46] But in a classic chicken-and-egg scenario, investments in full-scale production of advanced batteries are contingent on the market that would be created by the ZEV mandate, while the mandate is on hold pending the introduction of advanced batteries.[47] Perhaps CARB will stand by the postponed ZEV mandate, but history suggests that this is unlikely; without the mandate, the production of advanced batteries for use in electric vehicles will probably be delayed indefinitely.

If past experience tells us anything about the future, we should be wary of expectations and promised breakthroughs in battery technology. Without dramatic changes in prevailing energy prices or public policies, battery-powered electric vehicles will not be competitive with internal combustion cars on a head-to-head basis. It has been nearly a century since Edison first promised a better battery, but in many respects we are still waiting for it. Thus far, internal combustion technology has proved an elusive and fast-moving target. There is little reason to believe that even a significant federal battery technology program could catch up with, much less overtake, the enormous engineering accomplishments of the existing automobile industry.

But this conclusion does not mean that there are not legitimate uses for electric cars today—as fleet vehicles, as commuting cars, and as second cars for families that already own a gasoline automobile for long-distance travel. One partial solution might be to take electric cars and their batteries out of the development laboratory and put them into the hands of real drivers. Some will find the vehicles inadequate, but many others will not. By backing away from the early deadlines it set in the ZEV mandate, the California Air Resources Board has passed up the latest and perhaps best opportunity in recent history to establish a meaningful separate sphere for the electric vehicle.

7

TECHNOLOGICAL HYBRIDS AND THE AUTOMOBILE SYSTEM
From the Electrified Gas Car to the Electrification of the Automobile

> Every year the gas car becomes more electrical, and its electrical functions more important.
> Editorial, "The Electrified Gas Car," *Electric Vehicles*, 1917[1]

The specific relationship between the art and science of the storage battery, on the one hand, and expectations about the rate and direction of the advancement of the industry, on the other, produced both rising expectations and failed products, a pattern echoed by recent efforts to create a contemporary market niche for the electric vehicle. This pattern, however, begs the question: if internal combustion established a leading position over the electric vehicle and continued to increase the gap in performance, how could internal combustion ever be displaced? In other words, accepting that the dynamics of path-dependent technological choice create increasing advantages for the winning technology, can a winning technology ever be displaced; and if so, how?

Where does the electric vehicle stand in the larger context of the spread of electricity throughout society and the economy over the course of the twentieth century? By 1980 nearly 40 percent of all primary fuels were being used to generate electricity. Electricity has become the energy carrier of choice for almost every imaginable stationary application of power.[2] The failure of the stand-alone electric vehicle is an exception to the nearly global triumph of electricity. For instance, although the American rail network is not as electrified as European rail systems, diesel-electric locomotives burn diesel fuel to generate electricity that powers the driving wheels.[3] Beyond the perceived failure of the stand-alone electric vehicle, there are others ways to view the historical relationship between electrical and internal combustion power systems. Here, the concept and history of the hybrid vehicle serves as an entry point for analysis. In this chapter, after surveying a range of hybrid vehicles, we will consider the general issue of technological hybridization and its implication for the introduction of new technology into the automobile system.

THE STANDARD INTERNAL COMBUSTION AUTOMOBILE: A HYBRID VEHICLE

The logic of this chapter proceeds from a statement that at first glance may appear counterintuitive: the standard, gasoline-powered, internal combustion vehicle—the automobile, the universal artifact, the car in the driveway—is a hybrid gasoline-electric vehicle.[4] This claim will not sit well with those familiar with the technological history of the automobile industry since the 1970s. During the past several decades, hybrid vehicles have been viewed as one of the many possible alternatives to the standard internal combustion design that has dominated the industry for most of this century. Hybrids, along with electric vehicles and other alternative fuel vehicles, have been the subject of research and development and periodic concept cars; but in general, the hybrid has represented a possible automotive future, not the present or the past. Hybrid vehicles such as the Audi Duo, the Toyota Prius, and the Honda Insight are only now coming onto the market. Hybrids are the "next big thing," not the "last big thing." How can it be that the traditional car is already a hybrid?

Quite simply, what we know as the internal combustion automobile is a hybrid motor vehicle consisting of a relatively large internal combustion engine assisted by a small electric motor and a charging system for starting, lighting, and ignition. This design, which emerged in the 1910s with Kettering's development of the electric starter motor, combines important attributes of both electric and gasoline vehicles. In this sense, the contemporary standard American automobile is a specific type of technological hybrid: the electrified gasoline car. The success of this hybrid configuration marked the end of an intensive period of technological experimentation that began in the 1890s; up to the present day, this configuration continues to serve as the basis for the automobile.[5]

This experimentation implies that the electrified gasoline car was only one of several possible hybrid solutions to the problem of motorized road transportation. Other hybridization pathways were possible, both within the "black box" of the vehicle itself and at the larger system level. But because few of these alternative pathways have been seriously investigated, late twentieth-century observers of the automobile industry have failed to see the hybrid character of the modern motor car. Moreover, the engineering culture that grew up around nonhybrid (or "pure") technological alternatives such as the electric vehicle sharpened contrasts between existing conservative approaches and proposed radical technological changes. Electric enthusiasts sought to improve the electric-only alternative, while the mainstream industry turned inward, focusing on incremental extensions to the internal combustion trajectory. Like other assessments of the early history of the motor vehicle industry, this line of reasoning concludes that technological hybridization via the starter motor was an important part of the stabilization process that facilitated the dramatic expansion of the automobile industry. In other words, this argument does not deny the importance of Charles Kettering and the starter motor for the evolution of the universal vehicle. Nevertheless, other hybridization pathways were also available and more or less developed.[6]

COMBINATION ELECTRIC PATHWAYS

The first application of stored electricity for transportation involved motorized streetcars, and most early efforts to apply electricity for transportation focused on rail-based systems. Several types of battery-only streetcars were developed in Europe and the United States, while other installations combined trolley and battery technologies in a single system. Designed in part to address concerns about unsightly trolley wires in traditional city centers, these streetcars used the overhead wires to draw power and charge the batteries while traveling through suburban areas and then switched to battery-only mode while moving through the heart of the city.[7] Under this plan, outlying districts were served, while urban centers were spared the intrusion of overhead cables. In Hanover, Germany, site of an extensive system of combination battery- and trolley-fed streetcars, the municipal transit company operated no fewer than one hundred commercial delivery wagons with two sets of wheels: six small flanged ones that rode on rails and four large spoked ones for traveling on city streets. Described by *Electrical World* as "amphibious," these wagons could be loaded and towed to or from the street railway by a team of horses; and once on tracks, they could be towed by the streetcars.[8]

In the United States, where public opposition to overhead wires was weaker than in Europe, there was no need for such mixed systems. Electricity could be generated remotely at the station house, and the trolley quickly emerged as the preferred means for conducting electricity from busbar to motor. The successful electrification of American streetcars in the 1890s led to several schemes for taking electricity off the rails. The two most inventive ideas were the trackless trolley (later known as the electric bus) and the battery-powered hansom cab. Both combined existing technologies for electricity distribution and storage with pieces of established transportation systems. One early trackless trolley scheme envisioned privately owned electric carriages sharing a common overhead distribution system. This plan posed problems akin to allowing public access to railroads; for example, without an on-board secondary storage system, carriages could not pass each other or share a single set of overhead wires. An experimental double-trolley system, intended to operate "between the trolley car, which is a considerable and costly vehicle, and the electric automobile, which is . . . a machine entirely too expensive for public service," was built and tested in Issy, France, in 1900.[9] The trackless trolley hybrid was first used commercially in the United States in 1910 by the Los Angeles Pacific Railroad Company, which used the system to transport commuters from its streetcar lines to the Bungalow Land subdivision in Laurel Canyon (see figure 7.1). Before the installation of the trackless trolley, residents were ferried up and down the canyon roads by gasoline-powered coaches, but these proved unreliable and the required maintenance too excessive. The trackless trolley coaches promised "simplicity, reliability, economy in first cost, and low operating charges."[10] The service operated for eight years before being dismantled in 1918 after a fire destroyed the Lookout Mountain Inn, the mountaintop terminus of the line.[11]

The electric cab operation described in chapter 2 was also a hybrid system. The vehicles combined attributes of a horse coach, a battery-powered streetcar, and a bicycle. The design of the carriage and the rear-mounted position of the driver linked the vehicle to its horse-drawn predecessors. The battery (intended to be handled and exchanged as a single unit) was adapted from the streetcar schemes just described, while the large pneumatic tires were inspired by the use of high-pressure tires on bicycles. The hybrid nature of the elec-

FIGURE 7.1. First trackless trolley in the United States, Laurel Canyon, Los Angeles, California, 1910. *Source: R. W. Shoemaker, "The Trackless Trolley at Los Angeles," Electrical World 56 (October 27, 1910): 1002–3.*

tric cab was also reflected in the company's business organization. Although each unit within the holding company was an independent entity, the Metropolitan Street Railway Company combined electric cabs and trolleys under the ownership and control of a single group of investors.

THE ELECTRIFIED GAS CAR: ONE AMONG MANY HYBRID TECHNOLOGICAL TRAJECTORIES

As the first generation of primitive horseless carriages began to yield to slightly more complicated and evolved vehicles, new designs emerged that combined aspects of more than one type of primary propulsion system. Many of these combination products were stand-alone vehicles—that is, what today we would call hybrid vehicles. Nevertheless, the process of hybridization also occurred at higher organizational levels within the transportation system.

Hybrid Vehicles

By 1901 American George Fischer was exporting combination vehicles to replace horse-drawn buses in London. These vehicles were configured as what we would today call series hybrids: an all-electric drivetrain drew current from a lead-acid battery, which was, in turn, recharged by an internal combustion engine. Several were also put into service in New York City. H. L. Jespersen, the manager of the automobile department of Macy's department store, claimed that the Fischer truck was capable of running on batteries alone for up to two hours at a stretch.[12] By combining gasoline and electric drive systems in a single vehicle, this early hybrid attempted to achieve the best of both worlds, promising both the convenience and dependability of an electric with the range and refueling capability of a gasoline vehicle. Internal combustion powered the generator, so it could be easily and quickly refueled and would not need to be towed because of dead batteries. At the same time, the hybrid design sidestepped the thorny problem of a multiple-gear transmission and allowed the gasoline engine to run at or near its peak efficiency at all times. The risk, of course, was that the mixed system would deliver the worst of both worlds—the unreliability of a gasoline engine with the excess weight and limited range of an electric.

Fischer was neither the first nor the last supporter of the mixed internal combustion–electric drivetrain. Knight Neftal, manager of the

Boston station of the New England Electric Vehicle Company, developed one of the most exciting combinations of internal combustion and electric power. In 1902 he entered a gasoline auxiliary vehicle in a one-hundred-mile endurance test sponsored by the Automobile Club of America. Essentially, Neftal connected a stand-alone portable gasoline engine to a standard electric vehicle made by the Vehicle Equipment Company.[13] Later that year, his "special car" won the heat for electric vehicles in Narragansett, and in October he entered the vehicle in the long-distance run from New York to Boston and back. He had to withdraw after the first day, however, with an overheated engine.[14] Apparently, the auxiliary power source was not well matched with the vehicle and its batteries.

Several truck companies followed Fischer's example. The Couple-Gear Company, for instance, offered a line of commercial hybrid vehicles into the 1910s.[15] There were also sporadic reports of individual commercial hybrid vehicles being produced by companies such as the Champion Wagon Company of Oswego, New York; the Fuller Power Truck Company of Delphos, Ohio; and the Daimler Motor Company in England.[16]

In the passenger vehicle market, the Galt Motor Company of Ontario produced a hybrid in 1915 that claimed to deliver 23.5 miles per gallon. The manufacturer recommended it for use "where a power plant is needed since the electrical energy of the generator can be readily applied to numerous uses."[17] In 1916 the Woods Motor Vehicle Company of Chicago announced a hybrid car that would operate in either dual drive or electric-only mode and could supposedly travel forty-five miles per gallon; available features included regenerative coasting and braking and the ability to preset cruising speed (cruise control, as it is known today).[18] Meanwhile, Justus B. Entz, who had begun his career working for automotive pioneer Hiram Percy Maxim, developed a sophisticated electric transmission that was installed on a number of experimental vehicles produced by a merger of Baker, Rauch, and Lang in Cleveland with the Owen Magnetic Company of New York; General Electric also supported the venture. The Owen Magnetic vehicle used an infinitely adjustable electric slippage clutch in place of the traditional mechanical transmission, thereby obviating the need for a flywheel and a clutch.[19]

Despite the failure of these mixed vehicles to find a market, the interface between electric and internal combustion technology was both fluid and fertile. Designers, unable to foresee which combination of technological attributes would work best, experimented with different ways to use pieces of each technology to compensate for the weaknesses of the other. Neftal's approach was the simplest: run a stand-alone generator to provide electricity to charge the battery. The Owen Magnetic drive did not store electricity but used the magnetic clutch to minimize the problems associated with mechanical transmissions. The more complex hybrids such as the Woods dual power vehicle aimed to actually exceed the performance parameters of either of the independent component drive systems.

The story of the development of the starter motor is well known: Byron Carter's death in 1910 led Henry Leland to commission Charles Kettering to produce an electric starting system for the 1911 Cadillac.[20] But the preceding survey of other paths of hybridization indicates that the Kettering formula of connecting a small electric motor to an existing internal combustion engine was not the only technological configuration available. As early as 1902, *Horseless Age* columnist Albert Clough anticipated the development of the self-starter and noted that "in the combination vehicles . . . the problem of starting the engine is very beautifully solved by allowing the dynamo to act as motor, drawing from the battery and thus revolving the engine until it is in normal operation."[21] In short, early hybrid vehicles and the electric starter shared common inspiration; each, along with improvised designs such as Neftal's gasoline auxiliary, offered a different potential solution to the problems presented by the underlying component technologies. For whatever reasons, the Knight Neftal solution seems not to have provided sufficient range to allow the vehicle to tour with the rest of the Automobile Club of America. Similarly, the electric starter did not yield all of the benefits of the electric vehicle. In regular operation, the gasoline engine still must be geared to the driving wheel by a transmission and its speed regulated by a throttle. Nevertheless, the Kettering solution was good enough. It worked and allowed the electrified gasoline car to make significant inroads into the already narrowing market niche for pure electric vehicles.

Hybrid Transportation Systems

A broader view of the process of technological hybridization looks at the transport system as a whole. At the turn of the century, almost all observers of the nascent automobile industry predicted that each motive power—steam, gasoline, and electricity—would eventually find a place in the transportation system. In the words of *Horseless Age* contributing editor Albert Clough, "experience will furnish the decisive evidence which shall assign each motive power to its appropriate sphere."[22] The doctrine of separate spheres, had it stood the test of time, would have formalized a hybrid transportation system. Beyond the commercial realm, the early history of the automobile includes the emergence of several inchoate functional spheres that might have served as the basis for a hybrid system.

The turn-of-the-century household was the most obvious and durable agent of technological hybridization.[23] Given the high initial purchase cost of early automobiles, it is tempting to believe that consumers were forced into an irreversible choice among steam, gasoline, and electricity. But in fact the automobile was such a luxury that economy was not the issue; as the saying goes, "if you have to ask how much it costs to run it, you can't afford it." Most early car owners continued to maintain their horse and carriage and later added second or third vehicles. Moreover, with rapid structural and technological changes reverberating through the young industry, vehicles were often replaced every year or two. Testimonials in the pages of *Horseless Age* often reported drivers changing from one type of vehicle to another or using one type of vehicle while another sat in the stable. Vehicle registrations in Cleveland show that consumers frequently registered more than one type of vehicle at a time. Or an owner would register one make or style of vehicle first and then add a second type several weeks or months later.[24] In Paris a correspondent for *Electrical World* reported in 1901 that "many of the most prominent petroleum carriage users have one, two, or three electrically propelled carriages. . . . in a short time, every person in Paris of any means whatever will be the owner of at least an electric brougham, a Victoria, and a small petroleum voiturette."[25] In other words, the market was fluid. With various vehicles available to serve different needs, any number of hybrid patterns of vehicle use might have emerged.

Two representative usage patterns—the "doctor's car" and the "opera car"—demonstrate the range of outcomes inherent in the pre–World War I automobile system and suggest how consumer needs were pushing the automobile away from its roots as a toy for the wealthy toward becoming a useful product.

The first pattern coalesced around the need for dependable vehicle service for doctors and ambulances. By 1901 *Horseless Age* was printing a special section for medical practitioners, many of whom found that an electric vehicle was the best choice for emergency use. The electric was dependable, needed no time to warm up or saddle, and was quiet enough not to disturb sleeping neighbors in the middle of the night. For similar reasons, early ambulances were powered by electricity. But the same doctors who drove electric vehicles to make house calls on patients often drove steam or gasoline vehicles for their own personal touring.[26]

The second and longest lasting pattern of household hybridization assigned the steam or gasoline vehicle to the male driver and the electric vehicle to the female driver. Historian Virginia Scharff has argued that the electric vehicle was pawned off on female drivers. According to this view, the woman's opera car provided only limited mobility and enclosed its driver in the trappings of Victorian propriety. Crystal bud vases, card holders, and opera lights built into the electrics were no substitute for the power and license to range where one wished.[27] But this view may judge the electric vehicle of the 1910s too harshly. Society women might have welcomed increased range and performance; but perhaps the drivers of urban electric vehicles had no immediate need for these additional features, especially if they compromised the dependability, comfort, and ease of operation they had come to expect from their pure electrics. Moreover, the Scharff argument assumes that female motorists thought only in terms of individual mobility. But it is likely that most women who drove electric vehicles also had access to an internal combustion vehicle. In short, the hybrid household might have satisfied the transportation demands of all potential drivers.

In general, therefore, both the doctor's car and the opera car show that a range of different transportation needs existed and that they could be met by combining vehicle types. Whereas the early combination vehicles incorporated two prime movers into one artifact, the hybrid household of the 1910s saw the garage or stable as the black box and called on different technological alternatives depending on the specific transport service required. The hybrid household of the 1910s gradually disappeared, however, as gasoline cars became more similar to electric vehicles (i.e., enclosed bodies, easy to start, and comfortable to operate). The domestication of internal combustion gradually led consumers to own more than one type of internal combustion vehicle.

Hybridization also occurred at higher levels within the automotive system. For instance, at several points, state regulatory intervention was poised to create durable spheres of action by excluding some or all automobiles from specific parts of cities. Although many Americans were fascinated by and attracted to the automobile, the new contraption did encounter pockets of resistance. At the outset, the existing horse-based transportation sector, including livery owners, carriage makers, and teamsters, was not opposed to the use of the automobile per se. After all, the first horseless carriages that sputtered and wheezed their way through the nation's cities and towns hardly posed a tangible threat to the continued dominance of the horse. Rather, the horse-based sector viewed the horseless carriage and its operator as a nuisance. On occasion, a noisy vehicle or a reckless driver would frighten a team of horses or the teamster leading them, resulting in an overturned cart and perhaps a broken bone. In a few cases, opponents were able to enact regulations excluding automobiles from specific areas reserved for horse-drawn vehicles.

But the chaos that often resulted from motor races and public roads did not go unnoticed. Reporting on the Cosmopolitan road race in New York in 1896, an editorial in *Horseless Age* concluded that it was due only to "good luck" rather than "good judgment" that "no serious accidents resulted from this folly" and warned of the possible implications of a serious accident in future motor vehicle exhibitions on city streets: "Though no serious accidents were recorded on this occasion, it was tempting fate, and if persisted in will surely lead to disaster, and consequently to restrictive measures on the part of the authorities."[28]

By mid-1899 a number of American cities (including Baltimore, Chicago, New York, and San Francisco) banned automobiles from using the internal boulevards in urban parks on the grounds of public safety.[29] *Horseless Age* encouraged park board members to ride in an automobile before passing judgment and argued that they would quickly see that motor vehicles were readily controlled and posed no threat to pedestrians or horse-drawn carriages. In New York, police cited a small group of prominent citizens for violating the motor vehicle ban. The matter quickly ended up in court, and by December vehicles were allowed into Central Park but only after receiving a permit from the governing board. Permits were issued only for electric vehicles.[30] The following summer, park commissioner George C. Clausen granted special permits for four electric carriages to be placed in public service in New York City parks.[31] In another incident in

December 1900, a similarly discriminatory Baltimore ordinance was challenged when "a party of socialites" sent their electrics home and took their gasoline vehicles into Druid Hill Park to test the legality of the statute.[32] The issue was settled when the Maryland state legislature instituted a six miles per hour speed limit for motor vehicles operating within all city limits, thereby superseding the municipal law.[33] Efforts to keep horses and cars apart, even in limited domains, were probably destined to fail because motor vehicle drivers wanted to demonstrate the superior capabilities of the horseless carriage over the same terrain as, and in broad view of, the horse-drawn carriage users.

But what if the distinction between city cars and rural cars had been codified into law? Might such an arrangement have created a protected market niche for electrics and established a hybrid transportation system by fiat? Just as local fire codes banned all but electric vehicles from warehouses and loading docks, similar codes might have extended to the city limits and forced a shift in vehicle type at the city line or otherwise limited the number of internal combustion vehicles allowed in city centers. In 1914 *Electric Vehicles* reported that, among London-based automobile owners, it was "common practice . . . to use the gas car for touring the country and to depend on the public taxicabs for transportation within the metropolis. . . . [Gasoline car owners] when making visits to London by motor leave them in garages at the edge of town and employ taxicabs."[34] During World War I, Vienna considered a policy that would have denied licenses to all internal combustion vehicles and granted them only to electric vehicles.[35] Rural drivers in Italy were expected to drive on the right side of the road, while city dwellers drove on the left. When entering a city, at a certain point vehicles were forced to switch sides. These conventions persisted for the first several decades of the automobile era, and only a decree from Mussolini finally established an Italian standard of driving on the right.[36]

Alternately, what if the distinction between private and public transport had not come to be codified as rigorously and completely as it did? Had the car first emerged during a national crisis and the idea of jitneys (unregulated, privately operated cars that followed streetcar routes in search of would-be passengers) taken hold and become the norm instead of arriving late on the scene and serving merely to infuriate existing transit interests, drivers and passengers might have perceived the technology differently.[37] The success of the

Whitney plan to integrate rail and road transport service in New York (see chapter 2) could not have divided the transit interests because Whitney and his partners controlled both ventures. Without the deep and abiding personal attachment between owner and vehicle, perhaps other vehicle-sharing systems such as motor pools and cooperative delivery schemes would have found favor, resulting in different cultural valuations of automotive technology.

Admittedly, these systemic hybrid scenarios are little more than speculation; we know that internal combustion emerged as the dominant automotive technology. But the victor was, in fact, a specific type of technological hybrid that incorporated valued attributes of its major competitor—the electric vehicle. Moreover, identifying alternative systemic solutions to the problem of motorized transport clarifies the social and economic context from which the particular winning hybrid emerged; an easily controlled touring car was preferable to an urban electric vehicle with extended range. The winning technological hybrid—Kettering's small electric starter paired with a relatively large internal combustion engine—was not the only possible hybrid design available. Not only might a different artifact such as Knight Neftal's pure electric with a gasoline auxiliary have succeeded, but hybridization might have operated at a different level of aggregation within the transport system.

THE ELECTRIFICATION OF THE AUTOMOBILE SYSTEM

If the skeptical reader is still unwilling to accept the foregoing arguments about the hybrid character of the internal combustion automobile, consider the following fact: from the earliest days of the automobile industry, electrical components have played an increasingly important part in the functioning of the internal combustion vehicle. By the 1910s industry observers had recognized this trend and referred to the gasoline vehicle in editorials as "the electrified gas car." As early as 1913, *Electric Vehicles* noted that "the modern gasoline passenger vehicle contains and uses more distinctly electrical contrivances than any genuine electric vehicle ever designed." The Kettering self-starter was only the most recent addition to "well-appointed cars."[38] Every year manufacturers added more electrical features to internal combustion vehicles. In 1917 the Automotive Electric Association held its first annual meeting.[39] The following year, the National Automotive Electric Service Organization was created as an industrial organization for companies that produced electrical equipment for the automobile industry. All the while, numerous third-party manufacturers produced electrical accessories for the gasoline car, from steering wheel warmers to radiator blankets to lighted flag holders.[40]

In 1913 electrical components accounted for an estimated 1 percent of the total cost of an average vehicle. By 1920 the share had increased to 5 percent.[41] Ironically, even as the market niche for the electric vehicle was shrinking, the market for electrical components within the automobile industry was growing rapidly.

The dependence of the automobile system on electricity extended beyond the boundaries of the vehicle itself. Vehicle producers pioneered numerous innovations in manufacturing. The assembly line is the best known; but the automobile industry, as the vanguard of modern manufacturing, was a leading user of electricity in many ways. For example, the Ford Motor Company's massive electrical plant was chronicled in *Electrical World* in 1916, and separate reports described the extensive use of electric lighting to improve productivity at Ford.[42] The Goodyear factory in Akron and the Willys–Overland plant in Toledo were both early users of electric industrial trucks for moving materials around within their facilities.[43] The 1919 *Census of Manufactures* reported that electricity provided no less than 70.8 percent of the total primary horsepower employed in the automobile industry.[44] Meanwhile, outside the confines of the factory, the spread of traffic signals and electric street lights both increased the efficiency of cars in urban service and, in conjunction with new roadway designs, allowed high-speed driving at night.

Today it is inconceivable to think of a car without its associated electrical components; by any measure, the car is an electrified artifact. From accessories such as power windows, audio systems, defrosters, and halogen reading lights to critical monitoring and control functions now performed by onboard computer systems, the modern car is a veritable electrical cabinet on wheels. Manufacturers employ thousands of electrical engineers and purchase billions of dollars worth of components every year. Estimates of the relative contribution of electrical and electronic components to the total cost of producing a typical automobile range as high as 40 percent. Many new vehicle models simply cannot be operated if their various electrical systems and components fail. To remove electricity from the mod-

ern motor vehicle is impossible; the de-electrified gasoline car, to paraphrase Michael Schiffer, is useless. With at least a third of the total
cost of the car going to pay for electrical components, it is indeed
hard to claim that the modern automobile is not a technological hybrid.

HYBRID PATHWAYS RECONSIDERED

As was the case in chapter 6's discussion of technological expectations, the hybrid categories that emerged at the dawn of the automobile era also have parallels to the recent history of the automobile
industry. The different hybridization pathways that were present then
still exist but in a latent or dormant state. Although the electrified
gas car emerged as the universal standard automobile, other pathways,
although foreclosed, have each persisted in small market niches or
as innovations at the margins of the mainstream transportation
system.

Hybrid Vehicles

In addition to the electrified gasoline car, the past eight decades have
witnessed a host of other schemes for developing combination drive
systems. Several hark back to the Neftal gasoline auxiliary design of
1902. In the late 1970s, for instance, the manufacturer of the
Dynavolt electric car offered an auxiliary generator that could be
towed behind the vehicle for long-distance travel. Alan Cocconi, de-
signer of the AC drivetrain for the original General Motors Impact
electric vehicle that later became the EV1, adopted the same hybridization strategy for two cross-country trips in his most recent electric vehicle.[45]

The federal government has sponsored most of the post-1970 development of hybrid vehicles, much of it under the authority of the
Electric and Hybrid Research, Development, and Demonstration Act
of 1976 (see chapter 6). In 1974 the U.S. Environmental Protection
Agency tested a parallel hybrid Buick Skylark developed by engineer
Victor Wouk; and the Ford Motor Company built a handful of hybrid vehicles, including a full-size Econoline van, as a prototype for
a possible fleet-service vehicle.[46] Furthermore, the Electric Power Research Institute supported the development of the extended-range
electric vehicle (XREV), an electric service vehicle with a small engine generator to recharge the battery.[47] These official activities

tended to support research within the existing automobile paradigm. That is, in the United States the four leading domestic automobile manufacturers (later the Big Three) have been the dominant participants. Accordingly, these mainstream firms have tended to use new technologies to re-create the existing "universal" car. Recent product announcements from Toyota, Honda, and Audi also illustrate this point. Companies are introducing hybrid vehicles that are externally indistinguishable from existing internal combustion product lines. Most drivers might not even notice that the new vehicles represent an important point of departure from decades of automotive engineering practice. From a consumer acceptance and marketing perspective, these vehicles may be the best way to introduce fundamental technological change into the automobile industry.[48]

Hybrid Transportation Systems

Recent market research on California consumer preferences suggests a potential market for electric vehicles in a new generation of households. The hybrid household is already a fact of life for many families whose garages house roadsters and sedans, sport utility vehicles and minivans. All are still powered by internal combustion, but transportation researchers Thomas Turrentine and Kenneth Kurani have identified a small but significant group of consumers interested in adding limited-range electric vehicles to their households. This research focuses on a specific subset of the vehicle-buying public—new-car purchasers from households that already own an extended-range vehicle—and indicates that the California electric vehicle market might range from 98,000 to 213,000 vehicles (7 to 15 percent of the total number of new vehicles sold annually in the state and well in excess of the targets in the ZEV mandate). In part, the household would benefit from the convenience of recharging the vehicle at home and at other convenient locations, thereby obviating the need to use dedicated gas stations.[49] In all of these homes, the electric vehicle would complement the functions of an internal combustion vehicle; initially, relatively few internal combustion engines would be replaced because most of the displaced vehicles would be resold in the second-hand market. Even EV1, General Motors' highly touted entry into the consumer electric vehicle market, represents an implicit acceptance of the hybrid household strategy. By virtue of the target consumer demographics, very few of the buyers of this high-priced, two-seat sports car lack access to other types of vehicles for long-distance travel.

The past two decades have also witnessed renewed interest in the idea of appropriate spheres first proposed nearly a century ago. Many current transport policies tacitly support building hybrid transportation systems. First, companies, cities, transit districts, and state and regional governments have slowly discovered the benefits of tinkering at the margins with the standard automobile. Although other developed countries (such as Japan, France, and Italy) have long tolerated and even encouraged mini-cars and motorized tricycles, American regulators have held all vehicles to a single standard for safety and performance. Until recently, if a car could not survive on the interstate highway—regardless of whether or not it was intended for such service—it was deemed unfit. Such was the fate of the Citicar, the electric mini produced in the mid-1970s by Sebring-Vanguard. Although well received by drivers, the Citicar failed to pass a series of crash tests performed by Consumers Union in 1976, and soon the company was forced into bankruptcy.[50] Now public authorities have finally started to come around. New types of vehicles—station cars, neighborhood vehicles, community vehicles, and other limited-range or limited-performance designs—are being studied with an eye to how they might complement existing transportation resources without adding to total vehicle miles traveled.[51] Although these vehicles need not be powered by alternative fuels, their role makes them logical candidates for experimentation with electric and alternative fuel designs. Thus, public institutions may be stimulating the emergence of a hybrid transportation system by adjusting their regulatory and procurement policies.

In addition, public authorities have adopted specific regional strategies to mandate the diffusion of electric and alternative fuel vehicles. The best-known effort, California's Zero Emission Vehicle mandate, has been the subject of considerable debate. In the present context the mandate can be viewed as one attempt to set actual targets for the start of a hybrid transportation system within a given geographical area (that is, 10 percent of all new vehicles sold in California by 2003). If the California mandate goes forward, other regions may set similar targets. History has not erased the full range of hybridization pathways; however, creating a hybrid transportation system will obviously be more difficult now than it would have been in 1900.[52]

LOOKING FORWARD THROUGH THE PAST

Since the late 1960s, opponents of the continued expansion of the automobile system have struggled to articulate an alternative vision of motorized transport. The small, limited-range electric vehicle has been one component of that vision because the electric vehicle meets most, although not all, travel needs and is congruent with new cultural values. To paraphrase E. F. Schumacher, small could be beautiful. Consumers, however, have not chosen to act on this philosophy. Rather, after nearly three decades of wavering enthusiasm and support for the idea of electric vehicles, the total global market for road-ready electric passenger vehicles is still minuscule. Early sales reports from the major manufacturers that have introduced electric vehicles in California confirm this point: since the launch of General Motors' EV1 in the fall of 1996, fewer than 1,000 electric vehicles have actually been delivered; and Honda announced in early 1999 that it would stop selling its electric minivan, the EV Plus. For whatever combination of reasons—dissatisfaction with the available styles, dissatisfaction with range and performance, unwillingness to pay a premium price for an electric car—consumers have not been willing to accept the modern electric vehicle proposition.

Enter the hybrid vehicle—past, present, and future. This chapter has established that today's internal combustion vehicle is a technological hybrid that arose out of a specific set of historical circumstances and choices. This finding alone may be surprising, but accepting the hybrid character of the modern automobile may also foster incremental technological change within the automobile system. Reframing our ideas about the nature of the existing internal combustion vehicle may open our thinking to new pathways for the future. If we accept that the vehicle past and present is a hybrid, we should be able to expect the future to be similarly configured. The question then becomes not "if?" but "how?" How might a new hybrid differ from the old one? Whose hybrid will carry the day? Might the future be dominated by hybrid transportation systems as well as hybrid vehicles?

Any hybrid strategy would offer several important potential benefits. First, as I have mentioned in reference to the recent Audi and Toyota offerings, new hybrids can be virtually indistinguishable from their immediate predecessors. Consumers can be introduced to the hybrid proposition without disturbing any of their existing values and beliefs about the automobile. In fact, all new vehicles could be marketed as hybrid artifacts with different relative contributions of internal combustion and electrical energy depending on driving conditions and personal preferences. One purchaser might want to add extra batteries to an internal combustion design to allow all-electric operation when stopped in traffic, while another might opt for a different mix of auxiliary power and electrical energy storage. These options illustrate a second benefit: hybrids are flexible enough to allow for a wide range of consumer preferences. There could be performance hybrids, utility hybrids, and urban hybrids in addition to hybrids that simply reproduce the existing standard family sedan. This flexibility suggests a third possible benefit—the "microwave oven" effect. Over time, just as households have slowly come to appreciate the unique capabilities of microwave ovens, so, too, may drivers of hybrids gradually come to realize the benefits of using electric power without having to give up their cherished internal combustion vehicle. Thus, they may embrace the new values associated with the electrification of transportation. Eventually, these values may lead to more than incremental change and produce a transformed hybrid transportation system.

8

INDUSTRIAL ECOLOGY AND THE FUTURE OF THE AUTOMOBILE
Large-Scale Technological Systems, Technological Choice, and Public Policy

> It may now be taken as a fact that the introduction of the automobile has progressed far enough to make a sensible difference in the number of these disease bearing and annoying insects which are bred in certain residence localities of our cities and towns. . . . The automobile is blamed for all kinds of evil. When a good word can be said of it, let it be shouted from the housetops!
>
> Albert L. Clough, *Horseless Age*, 1909[1]

Conventional wisdom about the early history of the automobile has held that internal combustion was the best of the three technologies competing to dominate the nascent automobile market and that steam and electricity never really stood a chance. According to this view, internal combustion was the logical technology of choice by virtue of its intrinsic attributes: the high energy density of petroleum, the physical properties of prevailing construction materials, and the favorable power-weight ratio of the Otto four-cycle engine.

Without disputing the accuracy of any of these factual claims, this book has surveyed the early history of the automobile from a different standpoint: as part of the technological system for the provision of motorized road-based transport. I have tried to recapture the alternative technological configurations that might have been selected had events unfolded differently and to focus on the full range of social, economic, and technological factors that contributed to the resolution of the competition among steam, gasoline, and electric vehicles. My approach has also helped explain the challenges facing recent efforts to reintroduce electric vehicles into the American automobile industry. But while the first battle featured technologies at comparable levels of development, the skirmishes of the 1960s, 1970s,

and 1990s have pitted the electric alternative against a powerful embedded technology. In this chapter I suggest a framework for inter-
preting the long-term relationship between large-scale technological
systems and environmental change.

TECHNOLOGICAL CHOICE, SYSTEMS, AND SCALE

Building on recent studies in the field of industrial ecology, my model
uses three components—choice, system, and scale—to explain both
the state of current efforts to reintroduce an alternative to internal
combustion and the prospects for future systemic change.

Choice

Technological choice—the process by which one standard technology is selected from among a range of potential solutions to a given
problem or set of problems—is the first step in the creation of a
large-scale technological system composed of distributed and interdependent producers and users. Numerous factors besides the technical characteristics of the artifact affect the outcome of this process:
infrastructure, cultural norms, price signals, existing market interests
and organizations, and the attitude of the state. In the case of the automobile, these forces coalesced around the internal combustion engine
as the technology of choice so that, by the earliest years of the twentieth century, there was little doubt that internal combustion would
become the standard motive power for the private passenger vehicle.

The phenomenon of technological choice invites several observations. First, when one is uncertain about the long-term costs and
benefits associated with particular technological choices, it may seem
desirable to postpone choosing. If nothing else, delay allows time for
exploring potential consequences. Resolution is necessary, however,
for the emergence of a large-scale system. As we look back over a century of internal combustion and its widespread social and economic
costs, we may wish that the past had unfolded differently and that
the choice of prime mover for the automobile had been postponed.
What if all three technological options had developed together, each
with its own separate sphere of application, as many early industry
observers predicted? Would not today's choices be considerably less
costly and difficult? Perhaps, but the need to maintain overlapping
and duplicative infrastructures would have increased the costs associated with motorization; economies of production would have

evolved more slowly, as would diffusion of each of the technologies; and consumers would not have been able to reap as much social and economic benefit from automobility. In short, we may wish that early drivers had not chosen internal combustion as quickly and definitively as they did; but for the thousands who bought Ford's Model Ts and for the residents of urban America, the standard automobile could not come fast enough. The adoption of internal combustion as a clear and consistent technological standard oriented the process of system growth, directing search mechanisms (following Richard Nelson and Stanley Winter) and otherwise serving as a focusing device (following Nathan Rosenberg) that guided the future "sequence and timing of innovative activity."[2]

Second, technological choice often occurs quickly and outside the control of any single institutional actor. If the clock started ticking in November 1895 with the first American automobile race in Chicago, and if the internal combustion standard was assured by 1902, then the window of opportunity for influencing the outcome of the technological selection process was open for less than one decade.[3] In any technological endeavor that revolutionizes people's lives, it is remarkable how little attention people and institutions, both outside and within the nascent industry, can pay to the ramifications of the technological choice. Debate often focuses on short-term tradeoffs with little regard for long-term consequences. When it came to early automobiles, consumers and producers made informed judgments based on the best knowledge available, but they could no more anticipate the long-run social costs and benefits of their decisions in 1896 than we as individual consumers can today. Nor were public agencies any better equipped than other consumers to anticipate the unintended consequences of selecting one or the other technological alternative. Economic theorist and historian Paul David has described the position of large agents as the "Blind Giant's Quandary": dominant actors, like state policy-setting institutions, are most capable of influencing the course of future events "just when they least know what should be done."[4]

Third, although the preceding historical account made no mention of the environmental impacts of the various technological options, all three choices offered important environmental improvements for the manure-burdened cities of the late nineteenth century. The first cars dramatically improved the local environmental effects of transportation. This would have been true regardless of which technology emerged as victorious. On a per mile basis, the horse was and still is an extremely dirty and impractical technology for urban use. Historian Clay McShane has estimated that in the late 1890s horses passed between 800,000 and 1.3 million pounds of manure each day in New York City alone.[5] In general, cars took up less space on the street than a comparable horse and wagon; they consumed fuel only when they were moving, not when stationary; they were easier and less costly to store; they were less subject to exhaustion; they did not expose their passengers to any biological hazards; their waste did not pose a recognizable health threat. They were safer for passengers and pedestrians and spared urbanites the painful experience of watching "noble" horses suffer and die in the streets. In short, the first cars solved environmental problems.

The horse continued to play a role in the urban transport system into the 1920s, but the acute bottleneck that horse-based transport caused in late nineteenth-century American cities eased rapidly with the spread of the automobile.[6] The expanding automobile system gradually replaced the existing horse-based transport system and did the work of that system better and faster and with fewer environmental side effects, not incidentally allowing urban economic growth to continue. In the general case, technological choice can be seen as a response to prevailing social and environmental constraints. The negative side effects associated with technological selection are only apparent later, with the growth of a large technological system.

System

Technological systems emerge from a process of technological choice. A given technology or technological system is often neither inherently benign nor inherently harmful to society and the environment; rather, its context and growth trajectory establishes its cultural and environmental footprint. Doubtless some will argue that certain technologies are inherently damaging; but if we exclude a man-made Andromeda strain or any similar aberrant technological cataclysm, it is difficult to find an inherently dangerous artifact (for example, even one nuclear reactor poses a local, not a global, hazard). Given human nature and the frailty of global governance regimes, there are plenty of single technologies we would be wise to do without; but in the main, single instances of technology are relatively harmless. The technologies we most care about come as systems. Yes, we can measure the curb weight, the hauling capacity, and the engine displacement of a single car, but without the rest of the automotive transport system such a measurement is culturally devoid of meaning and practically useless. It is only in the context of an expansive technological system that encompasses a host of supporting subsystems that we can fully appreciate both the value and the meaning of this complex artifact. In *The Ecology of the Automobile*, Peter Freund and George Martin suggest the scope of the current automobile system: suppliers of parts, chemicals, fuels, roads, signals, repair sites, parking lots, traffic police, courts, insurance, scrap dealers, and various car lobbies and associations.[7] We must consider this expanded automobile system when we talk about the car's role in American society. More generally, it is the expansion of the technological system rather than any single artifact that places excess weight on specific social or cultural groups in society.

Paul David's Blind Giant's Quandary suggests an inherent structural limit on the ability to predict the negative consequences of systemic growth: we cannot anticipate the consequences of system expansion until initial technological selection has occurred; but once that initial choice has been made, certain degrees of freedom—certain pathways—become effectively foreclosed. Apparently, knowledge and experience vary inversely with the ability to intervene, and the growth of a system entails both reduction in uncertainty with a gradual decrease in tractability. In the case of the automobile, this phenomenon has been illustrated by researchers who have traced the expansion of the automobile using logistic growth models to describe a two-phase process of systemic development. The first growth pulse, which ended in the 1930s, was the replacement phase in which cars rapidly displaced horses. A second growth pulse, which began after World War II, corresponds to the full expansion of the automobile paradigm. This two-phase, or bilogistic, growth model helps explain the time and location of the environmental burdens conferred by the automobile.[8] Only as the automobile-based transportation system began to reach full scale, particularly in the Los Angeles basin in the mid-1950s, did citizens and public officials begin to perceive the negative environmental costs of that growth. The thick L.A. smog was the first warning sign, the beginning of the end of the unbridled expansion of the automobile. And although worldwide diffusion of the automobile has continued during the past forty years, the automotive bandwagon encountered its first barrier to continued growth with the public outcry in Los Angeles in the 1950s following the recognition that automotive exhausts were the principal source of eye-watering and asthma-exacerbating smog.[9]

Since that time, numerous additional constraints have emerged. The Environmental Protection Agency develops air pollution standards; the Department of Transportation administers fleet fuel efficiency (CAFE) regulations; the National Highway Traffic Safety Administration enforces safety laws first established by the Motor Vehicle Safety Act in 1966. The concentration of crude oil in the politically unstable Middle East has made energy dependence a further liability of our automobile culture, with overt costs ranging from the creation of the strategic petroleum reserve to military and political alliances with the Gulf States, the Persian Gulf War, and ongoing vexations with Saddam Hussein. In addition, tetraethyl lead, a fuel additive first used in the 1920s to prevent premature ignition and improve efficiency of combustion, was removed from gasoline in the United States in the late 1970s after it was shown to pose health risks, especially to children.[10]

Scale

Implicit in this discussion is a view in which new systems emerge as novel and clean technologies only to become gradually and inevitably dirty as they expand in scale. In their early phases, technologies such as the first automobile, the first commercial chlorofluorocarbons, and even the first nuclear reactors posed little or no environmental threat and may even have lessened the threat posed by the technologies they replace. But after 700 million vehicles are built or millions of tons of CFCs exhausted into the atmosphere, environmental carrying capacity becomes severely tested. Eventually, as the search for a solution advances, the process begins again. In this sense, the model is an outgrowth of the life-cycle methodology developed in the practice of industrial ecology, which applies some of the same methods and perspectives that natural ecologists have used to study traditional biological ecosystems to human social and economic systems: "Industrial ecology recognizes the unique role of humans in creating complex artifacts and institutions that force changes in materials and energy flows in both industrial and natural systems."[11] In their introductory text *Industrial Ecology*, Thomas Graedel and Braden Allenby connect a number of important technological innovations and their current negative environmental consequences to the initial problems they first solved (see table 8.1).[12] To that list of far-reaching innovations, we can now add the internal combustion–powered automobile. Like the other systems examined by Graedel and Allenby, the automobile's dramatic success has created a host of new social and environmental problems—from declining local air quality to excessive dependence on foreign supplies of crude oil to social fragmentation attributed to suburban sprawl.[13]

Dozens of other charges have also been heaped on the automobile, with critics blaming it for everything from aggressive behavior to noise pollution to creating an unhealthy sedentary population. Some of these pressures have resulted in important qualitative changes to the automobile system; cars today are safer, cleaner, and quieter. But if the expansion of the gasoline-powered automobile has been slowed, it has not stopped, especially if the international dimensions are taken into account.[14]

TECHNOLOGY, ENVIRONMENT, AND POLICYMAKING

Apparently, by the time we know enough to act, it is already too late to redirect the growth trajectory of emerging technological systems. What, then, are enlightened policymakers to do? Waiting for long-term social, technological, and economic evolution to solve problems is not an option. If it is possible to adopt a dynamic, long-term, systems view of the relationship between technology and environment, how does such a policy stance differ from one based on short-term concerns? Here, we explore several hypothetical policy goals in the context of the foregoing discussion.

Picking Near Winners

Although government direction of innovative activity runs afoul of some political ideologies and economic theories, support for second-best technologies (as proposed by David) is one possible response to the problem of the Blind Giant. This argument holds that because markets for emerging products select technological standards quickly *and* myopically, state policymakers should focus on picking near-winner or even "losing" technologies: alternatives that might, with modest and temporary support, offer longer-term social or environmental benefits. Over the long run, however, these policies are unlikely to alter the underlying dynamics of the perils of scale. Markets tend to converge on one or the other technological standard, making it increasingly hard to defend continuing support for the losers.

Table 8.1 *Historical examples of environmental problems resulting from industrial responses to prior problems or needs, showing technologies and systems following model of choice, growth, and impacts at scale*

YESTERDAY'S NEED	YESTERDAY'S SOLUTION	TODAY'S PROBLEM
Nontoxic, nonflammable refrigerants	Chlorofluorocarbons	Ozone depletion
Automobile engine knock	Tetraethyl lead	Lead in air and soil
Locusts, malaria	DDT	Adverse effects on birds and mammals
Agricultural fertilizer	Nitrogen and phosphorous fertilizers	Lake and estuary eutrophication
Overcoming horse-based limits on urban growth	Internal combustion automobile	Local air quality, unstable global markets

Source: Adapted from Thomas E. Graedel and Brad R. Allenby, *Industrial Ecology* (Englewood Cliffs, N.J.: Prentice Hall, 1995), 9.

Put another way, even if farsighted government planners had fought to preserve alternatives to internal combustion for one, two, or even three decades, it was not until well after World War II, almost fifty years after the winning technology had first emerged, that such policies would have borne fruit. In the interim, the nation had mobilized to fight two wars and the Great Depression. It is difficult to imagine support for, say, a hypothetical electric vehicle program surviving the budgetary demands of the Manhattan Engineering District. In the absence of perceptible risks from exclusive dependence on internal combustion, justification for an alternative technology program would eventually have evaporated. Thus, attempts to prevent premature technological selection may be justified on the basis of short- and medium-term efficiency concerns, but the long-run dynamics of technological and environmental change are unlikely to respond to efforts to manipulate the process of choice. The cause and effect are too far apart. But what we do not know *will* hurt us. Thus, the question appears to be "What will we choose to do when we figure it out?"

Strategic Niche Management

Several European scholars have proposed that the state use its institutional power to support and incubate high-risk experimental

technologies.[15] In this view, the state underwrites more than traditional activities such as research and development of radical technological solutions; it actively invests in sheltering innovations, thereby
enhancing the chances that new alternatives will diffuse and successfully displace entrenched systems. California's strategies of mandat-
ing new automotive technologies and subsidizing consortia to
promote the development and diffusion of new technology have been
cited as examples of strategic niche management. But what if the supported technology never establishes itself outside the protected domain? California's federal and state facilities, its electric utilities, its
universities, and its golf courses may all end up with the most advanced electric service vehicles available; but the scarce public funds
that paid for them might have yielded more social and environmental benefits if used differently. For example, the state might have purchased off-the-shelf service vehicles and used the money saved to
repair or replace its most polluting gasoline vehicles.[16] By limiting
government action, strategic niche management avoids some of the
problems associated with picking near winners but reproduces other
problems simply at a smaller scale.

THE NEXT-GENERATION VEHICLE

Given the previous argument, we can predict that sooner or later the
internal combustion regime system will fall victim to its own fantastic success and that from the mass of socio-technical constraints (technical efficiency, resource scarcity, supply vulnerability, concerns about
local air quality, limits on greenhouse gas emissions, and perhaps others still unknown) a range of new alternative technological solutions
will emerge and the process of technological selection, standardization, and system growth will begin again. Long-term technological
forecasting is by definition speculative, but we can draw some conclusions about how this process might unfold. Already, a wide range
of replacements for the traditional internal combustion–based vehicle
design exist in differing stages of development. The least radical alternatives (for instance, various applications of natural gas for
stand-alone vehicle transport) are already in use in many regions. Although this option is well adapted to existing infrastructure systems,
mobility patterns, and vehicle ownership, natural gas vehicles will
not address all the problems forcing change on the system. More radical solutions—such as electric vehicles, station cars and other car-shar-

ing schemes, and changes in urban structure (i.e., reintegration of working and living spaces)—may solve more problems but also face greater social inertia.[17] And even within a given category of alternative technology, a host of potential technological choices must still be made. In the choice of appropriate battery technology, for instance, we are seeing the same type of bottleneck that characterized the earlier competition among steam, gasoline, and electricity. None of the new batteries has emerged as the clear favorite, but many are poised on the brink of commercial introduction, even as groups continue to argue the merits of conductive versus inductive charging plugs. Technological choice, with all its attendant uncertainties, is the next logical step. The various communities that have fought to define alternatives to internal combustion will have to settle some of their own differences before the growth of a new technological system can commence.

The history of the technological competition among steam, gasoline, and electricity to power the American automobile and the technological and environmental changes that resulted suggest that our ability to anticipate and respond to the long-term negative impacts of large-scale technological systems is limited. Nevertheless, an expanding arsenal of analytic techniques may help us to better foresee future negative surprises resulting from the development and adoption of isolated technological fixes. Looking ahead, we can work to use our knowledge about the Blind Giant's Quandary and the internal dynamics of the process of technological selection to make better choices for the future. Yet at the same time we must always be prepared for the Law of Unintended Consequences.

EPILOGUE

I believe that the central station men should join in an active campaign with the electric vehicle manufacturers to try to do away with these "isolated plants." I think that every central station man should look on a stable, or any number of horses, as an isolated plant, and see to it that every possible endeavor is made to get the isolated plant work on the central station load.

Day Baker, General Vehicle Company, December 1913[1]

By examining the process through which internal combustion emerged as the standard motive power for the automobile and exploring a series of possible turning points in the development of the standard gasoline-powered vehicle, I have demonstrated that internal combustion was not the only alternative available to turn-of-the-century urban motorists. Had the Electric Vehicle Company succeeded in establishing dependable for-hire transport service, central stations might have recognized the potential of electric vehicles sooner than they did. Battery service might have been introduced ten years earlier; and with an expanding market for electric vehicle service, a progressive central station might have established remote battery exchange depots. It would certainly have been no more unreasonable to imagine battery exchange stations sprinkled throughout our neighborhoods than to envision gasoline stations equipped with underground tanks containing thousands of gallons of highly flammable refined petroleum delivered weekly by tanker truck. To be sure, in 1900, as the modern automotive transportation system was taking shape, electric vehicles were already at a disadvantage for touring. Charging infrastructure was underdeveloped, and battery exchange did not exist. An all-electric system might have been out of the question, but a hybrid system might have persisted for many years.

Nothing fails like failure. Following the collapse of the Electric Vehicle Company, internal combustion began to assume a dominant position in the developing motor vehicle market. Subsequent efforts, either by the Electric Vehicle Association of America or in the commercial vehicle realm, faced an increasingly uphill battle. The prospects for a hybrid outcome became more and more remote; and with the outbreak of World War I, the probability of long-term hybridization of the transportation system was effectively reduced to zero.

The intersystem rivalry that first emerged in the late nineteenth century continues today but obviously under very different circumstances. Eighty-five years later, Day Baker's call to do away with "isolated plants" still rings true. The horses are gone, however; it is the 200 million internal combustion engines that are the isolated plants belching forth wastes. For supporters of the contemporary battery electric vehicle, these modern isolated plants need to be electrified and put "on the central station load." EVAA president Frank Stone observed of the fleets of horses still traveling the streets of American cities in 1909: "We see hundreds of horse drawn vehicles which we are pretty certain *will not be seen* a few years hence." Today, too, must we assume that many of the internal combustion engines we see on the roads "will not be seen a few years hence." But will the replacement for the internal combustion engine be another internal combustion–based variant (a hybrid vehicle), or might it be something else (resulting in a hybrid system)? Within this context, the long-run prospects for increasing electrification of the national transportation system are good.

Public policy was not an important component of the process that produced the internal combustion–powered automobile, but it will likely be instrumental to the transition away from gasoline. The current round of electric vehicles does not appear to be sufficiently popular to convince regulators and automobile makers that a significant market exists for limited-range electrics. Perhaps the hybrid vehicle will serve as a transitional technology, while drivers and systems prepare for powered roadways or rediscover battery exchange. Perhaps one smog-choked city (such as Mexico City) or one oil-poor region (such as Japan) will be willing to incur the costs associated with switching technologies. Whoever dares to undertake this transition will do well to heed an essential lesson of the past: change will take more than a better battery. Following Hayden Eames, maybe the cheap, dense, and efficient fuel cell is the "accident" that we are all waiting for. But other events will surely occur; solving the problems posed by the success of internal combustion will require that we revisit the very definition of what constitutes an automobile. The alternative transport system will be more important than any single component technology.

Looking ahead, we can see no way to predict how today's technological choices will "bite back" in the future.[2] We cannot yet know, for instance, why we might, fifty years from now, want to unplug or at least refashion the family of information technologies known today as the Internet and the World Wide Web. Even electric vehicles, should they find widespread acceptance, are not a panacea. Interdependent regional electrical distribution systems are vulnerable to failure, just like any other technological system. And the more dependent we are on centrally generated electricity for transportation, the more disastrous the consequences of a serious blackout would be. In the end, therefore, technological diversity—encouraging and preserving a multiplicity of options—may be the best insurance against unwanted technological surprises. Although the cost of such a policy in terms of efficiency might be high, it is the price of living in a path-dependent world dominated by large technological systems.

Notes

1 .   RESCUING THE ALTERNATIVES TO INTERNAL COMBUSTION

1. The path dependent view of technological choice—the process by which a single standard technology is selected from among a range of potential technological options—contrasts with traditional economic theories of technological change in which the best technology is always readily apparent and always wins. For a recent overview of the phenomenon of path dependence, see Paul A. David, "Path Dependence and the Quest for Historical Economics: One More Chorus of the Ballad of QWERTY," *University of Oxford, Discussion Papers in Economic and Social History* 20 (November 1997): 1–48. Earlier works include W. Brian Arthur, "Competing Technologies: An Overview," in *Technical Change and Economic Theory*, ed. Giovanni Dosi et al. (New York: Pinter, 1988), 590–607; Paul A. David, "Clio and the Economics of QWERTY," *American Economic Review* 80 (1985): 332–37; and David, "Some New Standards for the Economics of Standardization in the Information Age," in *Economic Policy and Technological Performance*, ed. Partha Dasgupta and Paul Stoneman (New York: Cambridge University Press, 1987), 206–39.

2. This interpretation of technological system is derived from the work of Thomas P. Hughes, who for thirty years has enriched the historian of technology's theoretical vocabulary with his synthetic approach to large-scale technological systems and the seamless web of socio-technical problem solving. By studying the work of Thomas Edison and the creation of the electric power industry, Hughes was able to frame the technological system from the perspective of this entrepreneurial system builder. For Edison, the problems preventing the creation of a workable electrical distribution grid did not break down into discrete social and technical spheres. Instead, he faced a nexus of socio-technical barriers—a series of "reverse salients"—that he distilled into the critical problem of the filament for the incandescent light bulb. The filament bulb resolved a series of related social, technical, and economic problems. It reined in the stark, harsh light of the naked arc lamp for indoor use; it offered a reasonable balance between brightness and durability; and it allowed electric lighting to be priced competitively with existing gaslight systems. In Hughes's view, Edison succeeded because he was a "heterogeneous engineer" capable of devising solutions that simultaneously resolved technical and nontechnical bottlenecks to the growth of the system. Any study of the turn-of-the-century transportation system lacks a signal figure of Edison's stature; yet taken as a whole, transportation can still be analyzed through a systems lens.

    See Thomas P. Hughes, *Networks of Power: Electrification in Western Society, 1880–1930* (Baltimore: Johns Hopkins University Press, 1983); Hughes, "The Evolution of Large Technological Systems," in *The Social Construction of Technological Systems*, ed. Wiebe E. Bijker,

Thomas P. Hughes, and Trevor Pinch (Cambridge: MIT Press, 1987), 51–82; Hughes, "Techno-logical Momentum," in *Does Technology Drive History? The Dilemma of Technological Deter-minism,* ed. Merritt Roe Smith and Leo Marx (Cambridge: MIT Press, 1994), 101–14; and Bernward Joerges, *Large Technical Systems: Concepts and Issues* (Boulder, Colo.: Westview, 1988).

3. U.S. Bureau of the Census, *Abstract of the Census of Manufactures, 1914* (Washington: Gov-ernment Printing Office, 1917), 225.

4. James Rood Doolittle, *The Romance of the Automobile Industry* (New York: Klebold, 1916), v.

5. I base this observation on recent tours of Autoworld in Brussels and the Crawford Auto-Aviation Museum in Cleveland as well as childhood memories of the grand auto showroom at Harrah's in Sparks, Nevada.

6. James J. Flink, *America Adopts the Automobile, 1895–1910* (Cambridge: MIT Press, 1970), 307.

7. Ralph Nader, *Unsafe at Any Speed: The Designed-In Dangers of the American Automobile* (New York: Grossman, 1972). See also T. C. Barker, "Slow Progress: Forty Years of Motoring Re-search," *Journal of Transport History,* 3d series, 14 (September 1993): 142–65; and James J. Flink, "The Car Culture Revisited: Some Comments on the Recent Historiography of Auto-motive History," *Michigan Quarterly Review* 19 (1980): 772–81.

8. Josef Konvitz, Mark Rose, and Joel Tarr, "Technology and the City," *Technology and Culture* 20 (1990): 284–94; John A. Jakle and Keith A. Sculle, *The Gas Station in America* (Baltimore: Johns Hopkins University Press, 1994); Maxwell G. Lay, *Ways of the World: A History of the World's Roads and of the Vehicles That Used Them* (New Brunswick, N.J.: Jersey: Rutgers University Press, 1992); Clay McShane, *Down the Asphalt Path: The Automobile and the American City* (New York: Columbia University Press, 1994); Mark Foster, *From Streetcar to Superhighway: American City Planners and Urban Transportation, 1900–1940* (Philadelphia: Temple University Press, 1981); Clay McShane, "Urban Pathways: The Street and High-way, 1900–1940," in *Technology and the Rise of the Networked City in Europe and America,* ed. Joel A. Tarr and Gabriel Dupuy (Philadelphia: Temple University Press, 1988), 67–87; Bruce E. Seely, *Building the American Highway System: Engineers As Policy Makers* (Philadelphia: Temple University Press, 1987); Mark H. Rose, *Interstate: Express Highway Politics, 1939–1990,* (Knoxville: University of Tennessee Press, 1990 [1979]); and Kenneth T. Jackson, *Crabgrass Frontier: The Suburbanization of the United States* (New York: Oxford University Press, 1985).

9. Research on the relationship between urban structure and automobility was diverted by debate about the role of the automobile industry in the process of motorization. In a Senate staff report, Snell (1974) claimed that General Motors and its "auto-industrial al-lies" had purchased and sabotaged electric streetcar and interurban rail systems in Los Angeles and other cities and replaced them with their own diesel buses. Although brim-ming with errors, the Snell report was widely distributed, and its claims became the sub-ject of several monographs as well as the basis for a popular film, *Who Framed Roger Rabbit?*

Jackson (1985) has approvingly cited Snell's findings. Neverthless, refutations argue that Angelenos were happy to be liberated from the old, poorly maintained streetcar lines and that the actual fate of the city's public transit system was decided by local political consid-erations, not conspiratorial machinations in Detroit. Nonetheless, the debate drew atten-tion away from the deeper connections between the automobile and the city.

See Sy Adler, "The Transformation of the Pacific Electric Railway: Bradford Snell, Roger Rabbit, and the Politics of Transportation in Los Angeles," *Urban Affairs Quarterly* 27, no. 1 (1991): 51–86; Scott L. Bottles, *Los Angeles and the Automobile: The Making of the Modern City* (Berkeley: University of California Press, 1987); Bradford C. Snell, *American Ground Transport* (Washington, D.C.: U.S. Senate, Subcommittee on Antitrust and Monopoly of the Committee on Judiciary, 1974); and David J. St. Clair, *The Motorization of American Cities* (New York: Praeger, 1986).

For an exception that looks explicitly at the relationship between vehicle and city, see Martin Wachs and Margaret Crawford, eds., *The Car and The City* (Los Angeles: University of California Press, 1992).

10. Claude S. Fischer, *America Calling: A Social History of the Telephone to 1940* (Berkeley: Uni-versity of California Press, 1992), 8. Typical of this genre are Michael Berger's *The Devil Wagon in God's Country* (Hamden, Conn.: Archon, 1979) and Norman Moline's *Mobility and the Small Town, 1900–1930* (Chicago: University of Chicago, Department of Geogra-phy, 1971). Berger has traced the cultural path of the automobile through rural America, noting how initial resistance gave way to grudging acceptance and finally to widespread adoption. Moline's study examines the reaction of the town of Oregon, Illinois, to changes in "the friction of space" during the first thirty years of the twentieth century. Patterns of

sociability and mobility, concentration and deconcentration are evaluated in light of the changes accompanying the spread of the automobile.

Others, such as Howard Preston in *Automobile Age Atlanta: The Making of a Southern Metropolis, 1900–1935* (Athens: University of Georgia Press, 1979), have explored regional responses to the arrival of the automobile. In *Americans on the Road: From Autocamp to Motel, 1910–1945* (Baltimore: John Hopkins University Press, 1997 [1979]), Warren Belasco examines the motivations for and the effects of touring and car camping.

11. In his introduction to the proceedings of a conference commemorating the 1986 centennial of the automobile, Barker (1987) noted the limitations that the American experience has placed on the field and welcomed the identification of a host of alternative patterns of technological adoption. Motorcycles, for instance, dominated the motor vehicle landscape in England in the 1920s, while in Germany mass motorization before World War II took place via heavy trucking and buses. Other contributors to the centennial conference reported on late-blooming technological developments in countries such as Czechoslovakia and Zaïre. Beyond broadening general understanding of the social and economic consequences of the spread of the automobile, the variety of technological trajectories identified also suggest the need to revisit prevailing ideas about the American experience. Participants went so far as to question the conference's own existence by asking what precise event should mark the start of the automobile era. Significantly, however, Barker (1993) did not reexamine standard assumptions about the choice of motive power, attributing the success of internal combustion to "its advantageous power-to-weight ratio when the portable energy source as well as the engine itself were both taken into account" (146). Nevertheless, within the realm of automotive historians, his reframing of the process of motorization—international, polymorphic, and mutable—was a considerable methodological advance.

See Theo Barker, *The Economic and Social Effects of the Spread of Motor Vehicles: An International Centenary Tribute* (London: Macmillan, 1987); and Barker, "Slow Progress," 146.

12. Charles C. McLaughlin, "The Stanley Steamer: A Study in Unsuccessful Innovation," in *Explorations in Enterprise,* ed. Hugh Aitken (Cambridge: Harvard University Press, 1965), 259–72. Other histories of steam vehicles have also tended toward antiquarianism. See, for instance, John H. Bacon, *American Steam-Car Pioneers: A Scrapbook* (Exton, Pa.: Newcomen Society of the United States, 1984), who argues that the Stanleys borrowed extensively from earlier inventors such as Sylvester Roper, George Long, and George Whitney, as well as George Woodbury's fictional account, *The Story of a Stanley Steamer* (New York: Norton, 1950).

13. Andrew Jamison, *The Steam-Powered Automobile: An Answer to Air Pollution* (Bloomington: Indiana University Press, 1970); and Gary Levine, *The Car Solution: The Steam Engine Comes of Age* (New York: Horizon, 1974). McShane's *Down the Asphalt Path* stands out as a notable exception to this general trend. Although McShane dismisses the idea that steam "on common roads" could have substituted for the spread of the railroad, noting that the steam technology of the early nineteenth century was "clearly premature," he describes how later efforts to introduce steam road vehicles foundered on public distrust of the technology and fear of boiler explosions (83). The use of steam vehicles in the 1860s and 1870s was limited not by technological constraint but by local regulation (97–98). By the time trolleys and bicycles had established the desirability and feasibility of high-speed, street-level transport, internal combustion was already on the horizon. McShane concludes that although "steamers might have been more efficient than horses . . . , they were inferior to internal combustion cars." By integrating urban history with the histories of road building and car making, he has made an important contribution to the field but one that still perpetuates essential misconceptions about the relationship among steam, gasoline, and electric vehicles. In particular, McShane accepts without question the inevitability of a single, standard automobile, assuming that the electric option was doomed by "inferior electric technology" and that local governments and institutions welcomed all three types of vehicles indiscriminately (110).

14. Thomas P. Hughes, *Elmer Sperry: Inventor and Engineer* (Baltimore: Johns Hopkins University Press, 1971); Ronald Kline, *Steinmetz: Engineer and Socialist,* (Baltimore: Johns Hopkins University Press, 1992); and Harold L. Platt, *The Electric City: Energy and the Growth of the Chicago Area, 1880–1930,* (Chicago: University of Chicago Press, 1991).

15. Jackson, *Crabgrass Frontier,* chap. 6; Brian J. Cudahy, *Cash, Tokens, and Transfers* (New York: Fordham University Press, 1990), 41; Charles W. Cheape, *Moving the Masses* (Cambridge: Harvard University Press, 1980), 60–70; and Mark D. Hirsch, *William C. Whitney, Modern Warwick* (New York: Dodd, Mead, 1948), chap. 14.

16. James J. Flink, *The Automobile Age* (Cambridge: MIT Press, 1988); John B. Rae, "The Electric Vehicle Company: A Monopoly That Missed," *Business History Review* 29, no. 4 (1955): 298–311; Rae, *American Automobile Manufacturers: The First Forty Years* (Philadelphia: Chilton, 1959); and Rae, *The American Automobile Industry* (Boston: Twayne, 1984).

17. Richard Schallenberg's excellent *Bottled Energy* (Philadelphia: American Philosophical Society, 1982) was the earliest contribution to this literature. Although primarily writing a history of battery technology, Schallenberg saw electrical storage systems as a way of looking across utility, transit, and automotive applications of electric power—setting the stage for the current spate of studies.

   Volti (1990) and Schiffer (1994) pick up where Schallenberg left off. Volti attributes the outcome of the technological battle to the availability of complementary technologies such as starter motors, transmissions, and electric charging stations and connects the process of technological choice to the community of prospective users. His qualitative assessments of the various technologies emerge from the link he sees between the automobile and the interests of wealthy, white, predominantly male drivers in search of a plaything. Schiffer also integrates technological and social perspectives with considerable skill and enthusiasm, in the process providing the first historical periodization for the early history of the electric car. During the pioneering phase, from the middle 1890s until just after the turn of the century, the fundamental outlines of the American automobile were still unclear, whereas the electric revival in 1910 spawned the classic age, during which electrics briefly vied for a lasting but limited role in the provision of urban transport service. Although Schiffer's timing is not above reproach and his analysis overlooks the persistence of electric commercial vehicles, his periodization is a valuable and important starting point. See Rudi Volti, "Why Internal Combustion?" *Invention and Technology* (Fall 1990): 42–47; and Michael Schiffer, *Taking Charge* (Washington, D.C.: Smithsonian Institution Press, 1994)

   Continuing this line of research, Dutch scholar Gijs Mom has produced an impressive array of studies of the history of electric transportation, culminating in a Dutch dissertation that will be available in English as a companion study to the present volume (Johns Hopkins University Press, forthcoming). In addition to documenting the history of experiments with electric vehicles throughout Europe, Mom develops a compelling cultural theory to explain consumer preference for the internal combustion automobile as an adventure machine and concludes that even a high-performance electric would have failed to meet the demand for a challenging ride and a vehicle that allowed operators to demonstrate their skill and technological prowess. See Gijs Mom, "As Reliable As a Streetcar: European versus American Experiences with Early Electric Cars (1895–1914)" (paper presented at the SHOT Conference, Charlottesville, Va., October 1995); Mom, "Das Holzbrettchen in der schwarzen Kiste: Die Entwicklung des Elektromobilakkumulators bei und aus der Sicht der Accumulatorenfabrik AG (AFA) von 1902–1910" [The wooden thin plate in the black box: The development of the electric accumulator at and from the viewpoint of the AFA, 1902–1910], *Technikgeschichte* 63, no. 2 (1996): 119–51; and Mom, "The Technical Dilemma: Automotive Technology between the Petrol and the Electrical Vehicle (1880–1925)" (Ph.D. diss., Technical University of Eindhoven, 1997).

18. Merrit Roe Smith and Leo Marx, eds., *Does Technology Drive History? The Dilemma of Technological Determinism* (Cambridge: MIT Press, 1994); John M. Staudenmaier, *Technology's Storytellers: Reweaving the Human Fabric* (Cambridge: MIT Press and the Society for the History of Technology, 1985).

19. In its more extreme incarnations, this approach can be seen as arguing against the existence of physical, objective constraints imposed by technological artifacts and arguing for what Berger and Luckmann (1966) call "the social construction of reality." See Peter L. Berger and Thomas Luckmann, *The Social Construction of Reality: A Treatise in the Sociology of Knowledge* (Garden City, N.Y.: Doubleday, 1966).

20. Callon's (1980, 1987) actor-network theory, Latour's (1987, 1992) generalized symmetry postulate, and Pfaffenberger's (1990, 1992) anthropology of technology all represent different means of realizing similar ends—namely, privileging neither technology nor society in proximate explanations of social and technical change. See Michel Callon, "The State and Technical Innovation: A Case Study of the Electrical Vehicle in France," *Research Policy* 9 (1980): 358–76; Callon, "Society in the Making: The Study of Technology As a Tool for Sociological Analysis," in *The Social Construction of Technological Systems*, ed. Weibe E. Bijker et al. (Cambridge: MIT Press, 1987), 83–103; Bruno Latour, *Science in Action: How to Follow Scientists and Engineers through Society* (Cambridge: Harvard University Press, 1987); Latour, "Where Are the Missing Masses? The Sociology of a Few Mundane Artifacts," in

Shaping Technology/Building Society, ed. Wiebe E. Bijker and John Law (Cambridge: MIT Press, 1992), 225–58; Pierre Lemonnier, *Technological Choices: Transformation in Material Culture Since the Neolithic* (New York: Routledge, 1993); Bryan Pfaffenberger, "The Harsh Facts of Hydraulics: Technology and Society in Sri Lanka's Colonization Schemes," *Technology and Culture* 31, no. 3 (1990): 361–97; and Pfaffenberger, "Technological Dramas," *Science, Technology, and Human Values* 17, no. 3 (1992): 282–312.

21. See, for instance, Lay's *Ways of the World,* 197–201, on right- and left-side driving rules. Following the collapse of the Association of Licensed Automobile Manufacturers in 1911, the responsibilities of the standardization bureau were passed to the Society of Automobile Engineers, later renamed the Society of Automotive Engineers. See George V. Thompson, "Intercompany Technical Standardization in the Early American Automobile Industry," *Journal of Economic History* 14 (1954): 60–81.

On behavioral standards, see Lemonnier, *Technological Choices,* and McShane, *Down the Asphalt Path,* 173–202.

22. For sake of comparison, the current annual average is nearly 12,000 miles per vehicle. See U.S. Department of Transportation, Bureau of Transportation Statistics, *National Transportation Statistics, 1996* (Washington, D.C.: Government Printing Office, 1995), 34.

23. Jesse H. Ausubel and Hedy E. Sladovich, eds., *Technology and Environment* (Washington, D.C.: National Academy Press, 1989); and T. E. Graedel and B. R. Allenby, *Industrial Ecology* (Englewood Cliffs, N.J.: Prentice Hall, 1995).

---

2. WILLIAM C. WHITNEY, ALBERT A. POPE, RICHARD W. MEADE, AND THE ELECTRIC VEHICLE COMPANY

1. "Electric Vehicles," *Electrical World* 30, no. 7 (August 14, 1897): 182.
2. C. E. Corrigan, "Condition of the Horseless Carriage Industry," *Western Electrician,* January 1, 1898, p. 9.
3. "E. V. Company Placed in Receiver's Hands," *Automobile* 17 (December 12, 1907): 881.
4. John B. Rae, "The Electric Vehicle Company: A Monopoly That Missed," *Business History Review* 29 (December 1955): 311.
5. James J. Flink, *America Adopts the Automobile, 1895–1910* (Cambridge: MIT Press, 1970), 307; Michael Brian Schiffer, *Taking Charge: The Electric Automobile in America* (Washington, D.C.: Smithsonian Institution Press, 1994), 72–73. In *Bottled Energy: Electrical Engineering and the Evolution of Chemical Energy Storage* (Philadelphia: American Philosophical Society, 1982), 261, Richard Schallenberg has also concluded that the "complex exercise in incorporation was built upon a minimal technological base."
6. The two carriage bodies were produced by the Charles S. Caffrey Company of Camden, New Jersey, and the Crawford Wheel and Gear Company of Hagerstown, Maryland. See "Horseless Carriages," *Electrical Review* 27 (November 6, 1895): 265.
7. Through the early 1890s, American electrical engineers remained uncertain about the advisability of using stationary batteries. By 1895, however, with the creation of the expanded ESB, the industry was on firmer ground. The combined ESB consolidated all relevant "rights, patents and licenses" of Consolidated Electric Storage, General Electric, Electric Launch and Navigation Company, Brush Electric Company, and the Accumulator Company of Philadelphia. See "A Powerful Storage Battery Combination," *Electrical Review* 25 (December 12, 1894): 298.
8. On the early history of Morris and Salom's efforts, see Pedro G. Salom, "Automobile Vehicles," *Journal of the Franklin Institute* 141 (April 1896): 279–294; Salom, "The First Automobile Race in the United States," 1949, Western Reserve Historical Society, Cleveland; Ernest H. Wakefield, *History of the Electric Automobile: Battery-Only Powered Cars* (Warrendale, Pa.: Society of Automotive Engineers, 1994), 41–47; Schallenberg, *Bottled Energy,* 255, on work for the Julien and Consolidated Companies; Genevieve Wren, "Pedro G. Salom," in *The Automobile Industry, 1896–1920,* ed. George S. May (New York: Facts on File, 1989), 408–10; and Schiffer, *Taking Charge,* 45–48.
9. William Morrison, a Des Moines inventor, reportedly produced an electric surrey in 1891 and sold it to J. B. MacDonald, president of the American Battery Company in Chicago, for use at the Columbian Exposition (Wakefield, *History of the Electric Automobile,* 20). *Horseless Age* claimed that Fiske Warren, a Boston entrepreneur, built the first electric carriage ("The First American Electric Carriage," *Horseless Age* 3 [December 1898]: 18). Day Baker, an electric vehicle representative who was active in the Electric Vehicle and Central Station Association and later served a term as chairman of the Electric Vehicle Section of the National Electric Light Association, claimed that the first American electric vehicle

was built by F. M. Kimball in Boston in 1888 for P. W. Pratt. Later, Kimball was a member of the Electric Vehicle Club of Boston ("Mr. Day Baker, Lecturer," *Central Station* 9 [February 1911]: 172).

10. According to Russell H. Anderson, the fresh batteries had been preplaced. See his "First Automobile Race in America," *Journal of the Illinois State Historical Society* 47, no. 4 (1954): 353–54.

11. Salom, "Automobile Vehicles," 284–85.

12. H. F. Cuntz, "Story of the Selden Case and Hartford," 1940, Henry Cave Papers, National Automotive History Collection, Detroit Public Library, box 10, folder 28.

13. Hiram Percy Maxim, *Horseless Carriage Days* (New York: Dover, 1962 [1936]), 54

14. Henry Cave, "The Start of the U. S. Automobile Industry in Hartford, Connecticut," Cave Papers, box 4, folder 21.

15. Maxim, *Horseless Carriage Days,* 108–9

16. "An Electric Carriage Company," *Electrical Review* 28 (January 22, 1896): 53.

17. Schiffer, *Taking Charge,* 70. Wakefield suggests that Morris and Salom suffered a cash-flow crisis soon after their vehicle station opened and were forced to sell the business, although newspaper and periodical accounts of the inauguration of fleet service cite Rice and Gibbs as president and vice-president of the venture from the outset (*History of the Electric Automobile,* 41–56). For a contemporary account citing Rice and Gibbs, see "To Run Electric Cabs," *New York Times,* March 7, 1897, p. 10.

18. "Electric Motor Cab Service in New York City—II," *Electrical World* 30 (August 21, 1897): 216.

19. "Electric Carriages in New York City," *Electrical Review* 30 (February 24, 1897): 85; and "Electric Motor Cab Service in New York City—I," *Electrical World* 30 (August 14, 1897): 185.

20. R. A. Fleiss, "The New Station of the New York Electric Vehicle Transportation Company—I," *Electrical World* 37 (January 5, 1901): 5.

21. "Electric Carriages in New York City," 85; "Electric Motor Cab Service in New York City—II," 215—16; and Salom and Gibbs cited in "Interesting Storage Battery Discussion, II," *Western Electrician* 23 (September 3, 1898): 133.

22. As discussed later, the pneumatic tires used on the first-generation cabs were especially short-lived and prone to failure.

23. "Work of the Electric Hansom in New York," *Horseless Age* 2 (July 1897): 12; "To Run Electric Cabs," *New York Times,* March 7, 1897, pp. 10–11; Hugh Dolnar, "The Electric Vehicle Company," *American Machinist* 21 (September 29, 1898): 716. Based on two days of New York cab records in 1899 (1,251 paid out of 2,068.75 gross) and twelve months of New Jersey vehicle service in 1900–1901 (12,884.5 paid out of 19,244.5 gross), the actual figure is 66.23 percent paid. For New York data, see William F. D. Crane Papers New Jersey Historical Society, Newark, box 2, folder 7, For New Jersey data, see Frank C. Armstrong Papers, Henry Ford Museum and Greenfield Village Research Center, Dearborn, Mich., box 6, folder 2.

24. This statement refers only to direct operating expenses: forty cents per mile based on 12.5 miles per day per cab and sixty calls per month per cab. These figures do not include the cost of early battery replacement, as noted later. In addition to sources cited in note 23, see *Electrical World* 37 (January 5, 1901): 5; and *Electrical Review* 31 (November 3, 1897): 212.

25. See the certificate of incorporation, Cave Papers, box 5, folder 1. Initial capitalization was $10 million, a dramatic increase from the $10,000 capitalization of the Electric Carriage and Wagon Company.

26. James Bradley and Richard Langworth, "Calendar Year Production: 1896 to Date," in *The American Car Since 1775,* by Automobile Quarterly (New York: Dutton, 1971), 138.

27. Although no one has collected data on the number of homemade cars, the list of vehicle registrations in Detroit reports the brand of vehicle registered. As of 1904, a small number were listed as "own make." See "Automobile Directory of Detroit, Second Edition," National Automotive History Collection, Detroit Public Library.

28. Maxim, *Horseless Carriage Days,* 104–9.

29. "Articles of Agreement between Columbia Electric Vehicle Company and Studebaker Brothers, New York Branch, April 12, 1899," Crane Papers, box 2, folder 4; and "Agreement between Burr and Company, New York City, and Electric Vehicle Company, New Jersey, March 13, 1899," box 2, folder 8.

30. In 1902 W. H. Palmer admitted that the vehicles used between 1897 and 1899 "were equipped with various sizes of battery, and it was not until 1900 that an equipment was secured using a uniform size in all vehicles of the brougham and hansom types." See his

"The Storage Battery in the Commercial Operation of Electric Automobiles," *Electrical World* 39 (April 12, 1902): 646.

31. On Condict, see his biography in *Electrical World* 38, no. 24 (1901): 976. For a description of the Englewood and Chicago streetcar line, see *Western Electrician* 21 (August 7, 1897): 71–75. For Condict's appointment as chief engineer and his work on the central station, see *Western Electrician* 21 (December 25, 1897): 365; and "New Station of the Electric Vehicle Company," *Electrical World* 32 (September 3, 1898): 227–32.

32. Reports in contemporary engineering periodicals such as *Scientific American, American Electrician,* and *American Machinist* showcased the station as capital-intensive, automated, and standardized—in a word, modern. A lone crane operator, it was claimed, could grasp a discharged battery weighing almost a ton from the vehicle loading bay, place it on charge at one side of the room, and position the replacement battery for installation into the waiting cab. The cab could be back out on the street with a fresh battery in about three minutes. Pictures typically showed one employee in the cab of the crane poised high above the shop floor, surrounded by levers and cables. See "New York's Electric Cab Service," *American Electrician* (September 1898): 407–18; and "Central Station of the Electric Vehicle Company," *Horseless Age* 3 (September 1898): 9–17. Despite the impressive capital goods arrayed in support of the cab service, the operation was still very labor intensive, depending on both skilled and unskilled labor around the clock.

33. William F. D. Crane to George H. Condict, October 11, 1898, Crane Papers, box 1, folder 2.

34. See his technical notebook dated February–April 1899, Crane Papers, box 2, folder 3.

35. Fred Jenkins, in a speech to the Association of Edison Illuminating Companies in 1899 titled "The Electric Automobile," noted that "one of the most serious and expensive defects in the present construction of the electric vehicle exists in the tires, the cost of repairs on tires in cab service often times exceeding that of any other operating expense." Jenkins also noted, however, extensive experimentation and anticipated improvements in this area (reprinted in Minutes, Association of Edison Illuminating Companies, 1899, 197–206). Gijs Mom, looking at the maintenance costs associated with Dutch and German electric cab fleets, has concluded that replacement of tires was the largest single item in the overall maintenance of electric cabs. Part of the explanation of the high cost of tire maintenance in the European fleets can be attributed to the relatively higher cost of rubber in Europe at the time, but Mom's general observation applies equally to the American situation: tires were an early and expensive problem for the electric cabs. Gijs Mom, "Das Holzbrettchen in der schwarzen Kiste: Die Entwicklung des Elektromobilakkumulators bei und aus der Sicht der Accumulatorenfabrik AG (AFA) von 1902–1910" [The wooden thin plate in the black box: The development of the electric accumulator at and from the viewpoint of the AFA, 1902–1910], *Technikgeschichte* 63, no. 2 (1996): 119–51.

36. "Specifications for Altering and Repairing the B. Brougham of the New York Electric Vehicle Transportation Company," May 17, 1900, Crane Papers, box 2, folder 7.

37. George H. Condict to Walter H. Johnson, December 16, 1899, Crane Papers, box 2, folder 8.

38. Pedro G. Salom, quoted in "Interesting Storage Battery Discussion, II," 133.

39. "Electric Automobiles," *Electrical World* 32 (November 5, 1898): 465–66.

40. Of teaching first-time drivers how to operate his early gasoline-powered vehicles, Maxim wrote: "before long anyone who forgot to push down the clutch and rasped his gears was in disgrace and considered not competent to drive. Before this, when someone rasped the gears, it was considered evidence that my design was faulty. From this time on there was no general criticism of the control system of the gasoline carriage" (*Horseless Carriage Days,* 131). This statement described the Mark VIII Lot 4 vehicles, dating his observation to 1899.

41. See, for example, *Horseless Age* editorials (discussed later). The magazine's correspondents tended not to take such an extreme negative view; for example, W. D. Hutchinson wrote:
    the electric automobile is stately, elegant, dignified—a gentleman's carriage, free from disagreeable noises, odors, or escaping vapors or fumes, and as pleasant and simple to operate as need be. In fact, so little technical skill is required that any man, woman or youth can run it. . . . With reasonable care and skill in handling the amount of service that can be obtained from this vehicle seems prodigious. I almost believe one could be run till the end of time and then be in fairly good shape, if no accident happened. On the other hand, the battery situation should be clearly understood by the intending purchaser, for right here we come up against a serious drawback, so far as I am able to see. Unlike the vehicle itself, the battery does deteriorate, and that rapidly, consequently is expensive to maintain, if a satisfactory capacity is to be obtained, introducing an expense item that is very considerable, as well as a source of care that is almost constantly present if heavy duty is required from it.

In contrast to the journal's editorial screeds, Hutchinson accepted the benefits and costs associated with using an electric vehicle ("Communications," *Horseless Age* 6 (April 4, 1900): 18.

42. Armstrong filed several complaints about the company's vehicles. In July 1905 he wrote to EVC management in Hartford, saying: "our Electric Vehicle Department is apparently being neglected. . . . We have been looked upon as absolute leaders in the manufacture of electric vehicles, but we are rapidly losing this reputation." By the end of the year, a frustrated Armstrong tried to go over the heads of management to Thomas F. Ryan, an original partner of William C. Whitney, and to Whitney's son, Harry Payne Whitney. Armstrong noted that monthly sales were down from earlier years and argued that "the direct cause of this [shortfall] is the failure of the Company management to use proper business foresight" (Frank C. Armstrong to Electric Vehicle Company, July 8, 1905, Armstrong Papers, box 1, folder 4; Armstrong to Harry Payne Whitney, December 15, 1905, box 1, folder 5; Armstrong to Thomas F. Ryan, January 4, 1906, box 4, folder 1).

43. Reprinted from the *New York Journal* in "A Chappie and a Horseman Try the New Horseless Carriage," *Horseless Age* 2 (March 1897): 15–16.

44. *Electrical Review* 36 (March 28, 1900): 306; and "Horseless Look of Our Automobiles," *Electrical Review* 37 (August 22, 1900): 185.

45. "Automobile Ice Wagons," *Electrical Engineer* 33 (January 12, 1898): 63.

46. "Snow and Mud Pictures," *Horseless Age* 3 (December 1898): 14.

47. "Electric Vehicles in the Recent Storm," *Electrical World* 32 (December 17, 1898): 657; and the *New York Tribune* account reported in "Electric Cabs Best in a Storm," *Western Electrician* 23 (December 3, 1898): 315.

48. "Electric Vehicles in the Recent Storm," 657.

49. "Who Shall Build the Automobiles?" *Electrical Engineer* 28 (January 26, 1899): 114; "New York Electrical Society—194th Meeting—at the Station of the Electric Vehicle Company, February 14th, 1899," *Electrical Engineer* 27 (February 23, 1899): 232; and "Travel during the Blizzard," *Electrical Engineer* 27 (February 23, 1899): 225.

50. Editorial, *Western Electrician* 25 (July 22, 1899): 50.

51. As of February 14, the EVC claimed to have eighty-four vehicles in service; see "New York Electrical Society—194th Meeting," 232.

52. Maxim, *Horseless Carriage Days*, 165–66.

53. "Snow and Mud Pictures," 14.

54. For his crime William Dollard spent two weeks in jail; see *Electrical World* 36 (October 29, 1900): 664.

55. See minutes of EVC executive committee meeting, August 28, 1899, Armstrong Papers, box 1, folder 6.

56. Gorman Gilbert and Robert E. Samuels, *The Taxicab: An Urban Transportation Survivor* (Chapel Hill: University of North Carolina Press, 1982), 34–35. Also see 1906 articles on introduction of mechanical fare measuring devices in *Commercial Vehicle, Power Wagon,* and *Horseless Age.*

57. "The Horse Must Go: Hotel Men Give Their Views on the Adoption of the Automobile," *New York Times,* October 31, 1899, p. 8.

58. See, for instance, the letter from Thomas J. Regan to William C. Whitney, July 31, 1900, in which Regan concludes: "Mr. Day presented a comparative statement at the Committee Meeting today of the earnings, incomes, etc. of the Storage Battery Co for the six months January to June inclusive. It was so good I thought you would like to see it, so send it herewith." The report showed that 17 percent of sales ($242,383 out of $1,396,817) were derived from vehicle batteries. See William C. Whitney Papers, Library of Congress, box 100, letterbooks, series 2, vol. 11, July–October 1900.

59. "More Electric Cabs for New York," *Horseless Age* 3 (February 1899): 13; Westinghouse Electric and Manufacturing Company to Electric Vehicle Company, November 17, 1897, Crane Papers, box 2, folder 8.

60. "Rumored Automobile Trust" and "Trust in Automobiles," *New York Times,* February 22, 1899, pp. 2, 10.

61. "$25,000,000 Electric Vehicle Company Incorporated," *Horseless Age* 3 (March 1899): 7; and "The New Electric Vehicle Company," *New York Times,* February 25, 1899, pp. 1–7. There are conflicting accounts of changes in the capital structure of the EVC; it is possible that more than doubling the capitalization was a defensive move intended to make it harder for the company to be acquired by outside interests. But it seems more likely that the *New York Times* reported that the EVC had increased its capitalization when, in fact,

the New York Electric Vehicle and Transportation Company (the future regional operating company for the EVC) was established at this time.

62. Among other names, the group was also referred to as the Whitney-Elkins syndicate and the Whitney-Widener-Elkins syndicate. See "Automobiles," *Western Electrician* 25 (August 19, 1899): 112; (October 7, 1899): 199. On the general history of the Metropolitan Street Railway, see Charles W. Cheape, *Moving the Masses: Urban Public Transit in New York, Boston, and Philadelphia, 1880–1912* (Cambridge: Harvard University Press, 1980), 40–70.

63. Schallenberg, *Bottled Energy*, 225–39. On the general behavior of the Metropolitan, see Joseph P. Sullivan, "From Municipal Ownership to Regulation: Municipal Utility Reform in New York City, 1880–1907" (Ph.D. diss., Rutgers University, 1995); and Mark D. Hirsch, *William C. Whitney: Modern Warwick* (New York: Dodd, Mead, 1948).

64. The new conduit system of the Fourth and Madison Avenue line began operation in late 1897. Based on the success of the new lines and "public appreciation of the new system," MSR president H. H. Vreeland accelerated the schedule of electrification. See "The Underground Trolley in New York City," *Electrical World* 30 (August 28, 1897): 239–41; and "New York Notes," *Western Electrician* 22 (January 15, 1898): 42; 23 (September 17, 1898): 166.

65. The New York Gas and Electric Light and Power Company was organized in November 1898 by Whitney and Anthony Brady, among others, to consolidate control over New York's disparate light and power companies. See "New York Notes," *Western Electrician* 23 (October 15, 1898): 222; 26 (January 13, 1900): 34.

66. Hirsch, *William C. Whitney*, chap. 14.

67. Changes in the legal and political context for transportation service, especially the prospects for the New York subway, may also have shaped Whitney's thinking about the EVC; Governor Theodore Roosevelt and the New York State legislature had recently limited claims to perpetual franchise rights. For general discussion of evolution of transport franchises, see Cheape, *Moving the Masses*, 62–70; and Harry James Carman, *The Street Surface Railway Franchises of New York City* (New York: AMS, 1969 [1919]), chap. 8.

68. Cuntz, "Story of the Selden Case."

69. "Financial Development of the Automobile Industry," *Electrical World* 33 (May 13, 1900): 594.

70. John B. Rae, "Albert Augustus Pope," in *The Automobile Industry, 1896–1920* ed. George May (New York: Facts on File, 1989), 397.

71. "Whitney in a Big Automobile Trust," *New York Herald*, April 26, 1899, and "Millions to Back the Motor Trust," *New York Herald*, April 27, 1899, in Whitney Papers, box 191, clippings, November 28, 1898–April 27, 1899.

72. "Electric Vehicle Companies for Every State and Territory," *Horseless Age* 4 (May 24, 1899): 11.

73. "A $200,000,000 Enterprise," *Horseless Age* 4 (May 10, 1899): 6.

74. "Financial Development of the Automobile Industry," 594.

75. "The Automobile Ride," *Electrical World* 33 (June 3, 1899): 741; "The Great Electric Automobile Parade," *Electrical Review* 34 (May 31, 1899): 339; and "Automobile Ride of the National Electric Light Association," *Electrical World* 33 (June 3, 1899): 763.

76. The $75,000,000 Anglo-American Rapid Vehicle Company was reportedly incorporated in November 1899 "to combine English and American automobile companies." See "Automobiles," *Western Electrician* 25 (November 18, 1899): 304. On Croker and the New York Auto-Truck Company, see "Plans of the New York Autotruck Co.," *Electrical Engineer* 27 (January 5, 1899): 24; "Auto-Mobile, -Truck or -Wain," *Electrical Engineer* 27 (January 5, 1899): 27; "A Horseless City," *Electrical Engineer* 27 (January 19, 1899): 87; and "Industrial Combinations," *Western Electrician* 25 (August 19, 1899): 112. The General Carriage Company was reported to be interested in compressed air, among other prime movers. The Whitney interests, meanwhile, had already invested in compressed air through the American Air Power Company (later reportedly merged with the Compressed Air Company), which operated air-powered streetcars on the crosstown street car route at 28th and 29th streets. See letter from Thomas J. Regan to G. S. Hawkins, April 27, 1899, reporting that work on compressed air "is being pushed with all rapidity. . . . The motor has in the opinion of the Street R'y officers, passed its experimental stage, and the 28th and 29th Street proposition now means business" (Whitney Papers, box 99, letterbooks, series 2, vol. 6, December 1898–May 1899). On the announcement of General Carriage, see *Electrical Review* 36 (January 24, 1900): 98. The Philadelphia and New York traction interests were also reported to be "largely interested" in the American Automobile Company, a gasoline vehicle manufacturer. See "Automobile Notes," *Electrical Review* 35 (September 6, 1899): 156.

77. Maxim, *Horseless Carriage Days*, 165.
78. "Motor Vehicle Trust," *Horseless Age* 4 (May 3, 1899): 5.
79. Cuntz, "Story of the Selden Case."
80. "Minor Mention," *Horseless Age* 4 (June 7, 1899): 13; "$8,000,000 in Vehicles," *New York Times*, July 13, 1899, pp. 14–15; and minutes of EVC executive committee meeting, July 25–27, 1899, Armstrong Papers, box 1, folder 6.
81. "Manufacture of Electric Automobiles," *Electrical World* 35 (January 13, 1900): 53–56; and "Methods of Automobile Manufacture," *Electrical World* 35 (May 26, 1900): 799 801.
82. "Chicago Electric Vehicles," *New York Times*, May 5, 1899, pp. 3–4; and "New Automobile Companies," *New York Times*, May 6, 1899, pp. 2–3. Note that this interpretation varies from Hounshell's view of the Pope Manufacturing Company as an important conduit for the transfer of armory practice to the automobile industry. In my opinion, the imperative to acquire third-party producers to control competition edged out the progressive impulses that might have been developed in Hartford. See David A. Hounshell, *From the American System to Mass Production, 1800–1932: The Development of Manufacturing Technology in the United States* (Baltimore: Johns Hopkins University Press, 1984), chap. 5.
83. The Riker plant in Elizabethport, New Jersey, for instance, was repeatedly sought by the EVC. Riker resisted these offers until December 1900, and an unexplained fire damaged the factory less than a month later. See "The Electric Cab Experiments," *Horseless Age* 4 (August 30, 1899): 7; "Electric Vehicle Co. Absorbs Riker Company," *Horseless Age* 7 (December 12, 1900): 15; and "Minor Mention," *Horseless Age* 7 (January 9, 1901): 32.
84. See minutes of EVC executive committee meeting, December 26, 1899, Armstrong Papers, box 1, folder 6.
85. "London Notes," *Horseless Age* 4 (August 16, 1899): 15.
86. "No More Electric 'Fiacres' in Paris," *Horseless Age* 4 (August 30, 1899): 19.
87. "Automobile Club's Storage Battery Tests," *Horseless Age* 4 (August 30, 1899): 19.
88. "The Electric Cab Experiments," *Horseless Age* 4 (August 30, 1899): 7.
89. "Declaration of Independence," *Horseless Age* 5 (October 4, 1899): 5.
90. "Electric Vehicles and Their Limitations," *Horseless Age* 4 (September 27, 1899): 5.
91. "Lead Cab Funerals," *Horseless Age* 5 (October 25, 1899): 7.
92. "Elihu Thomson Favors Steam," *Horseless Age* 5 (October 4, 1899): 7. Thomson had experimented with "various elements, such as gas and gasoline," but had initially settled on an electric vehicle. *Horseless Age* interpreted his change of heart as a particularly significant event. See "New Horseless Carriage," *Electrical Review* 30 (June 30, 1897): 317.
93. "A Discredited Expert," *Horseless Age* 5 (October 18, 1899): 5–6.
94. "Storage Battery Financiering," *Horseless Age* 5 (October 18, 1899): 6–7.
95. "If," *Horseless Age* 5 (October 18, 1899): 7.
96. "Résumé," *Horseless Age* 5 (January 24, 1900): 9.
97. "Electric Vehicle Number Later," *Horseless Age* 5 (January 10, 1900): 8; emphasis added.
98. "The Autoelectrophobe," *Motor Age* 2 (March 29, 1900): 77.
99. "Electric Cabs," *Electrical World* 34 (September 9, 1899): 370; and "Electric Automobiles," *Electrical World* 34 (October 14, 1899): 566.
100. "A Plea for Discussion," *Horseless Age* 5 (March 28, 1900): 9.
101. The Chicago Automobile Transportation Company, operating electric stages and cabs, was announced in February 1901, and William Weedon was granted a permit to operate electric vehicles in Druid Park, Baltimore, in May 1901. See "Minor Mention," *Horseless Age* 7 (February 20, 1901): 30; 8 (May 8, 1901): 137.
102. "General Carriage," *Electrical World* 35 (January 6, 1900): 37.
103. "Lead Cab Finale," *Horseless Age* 5 (December 20, 1899): 8.
104. See A. L. Stevens to EVC, New York, December 19, 1899, Crane Papers, box 2, folder 8. The fire was also reported in "New York Notes," *Western Electrician* 25 (December 23, 1899): 372. Damage was estimated at $60,000.
105. George H. Condict to Walter H. Johnson, November 27, 1899, Crane Papers, box 2, folder 8. Condict was relatively sanguine about the prospects for the new vehicles: "Having gone through a similar experience with the first lot of cabs built by the company, we were not at all disheartened by these difficulties, although feeling chagrined at the fact that the vehicles did not show up better."
106. Even if the Whitney syndicate was active in the General Carriage Company, the business conditions for the EVC had not changed.
107. Jenkins, "The Electric Automobile"; and "Automobile News," *Electrical Review* 35 (August 16, 1899): 108.

108. Spencer Crane, "The Newport, R.I. Electrical Automobile Station," *Electrical World* 34 (December 16, 1899): 927–31.
109. William P. Kennedy to EVC, New York, November 21, 1899; and George H. Condict to Walter H. Johnson, November 27, 1899, Crane Papers, box 2, folder 8.
110. Ibid.
111. This is based on an estimated leasing charge of $150 per month, a four-month rental season, and seventy-five vehicles in service. See *Electrical World* 36 (October 27, 1900): 662.
112. "New England Electric Vehicle," *Electrical World* 35 (June 9, 1900): 888; "Automobile Charging Stations," *Electrical World* 35 (June 16, 1900): 891; and "Electric Automobiles in Boston," *Electrical World* 35 (June 16, 1900): 895–97. Additional stations were planned in Lynn, Salem, Waltham, and Lexington.
113. "Electric Automobiles in New England," *Electrical World* 37 (January 12, 1901): 136.
114. Re-creating an accurate financial picture of the operating companies is nearly impossible. Based on disparate reports, the statement regarding profitability seems accurate. Income from October 1900 to March 1901 is estimated at $4,690 greater than expenses, with a gross margin of slightly less than 30 percent. Reported revenues increased from $100,000 in late 1900 to $116,386 in March 1901, while reported expenses only increased $11,696 during the same period. See "New England Electric Vehicle Transportation Company," *Commercial and Financial Chronicle* 70 (April 7, 1900): 685; "New England Electric Vehicle Transportation Company," *Electrical World* 35 (April 14, 1900): 570; "Automobiles in Cities," *Electrical World* 36 (August 25, 1900): 288; "New England Electric Vehicle," *Electrical World* 36 (October 13, 1900): 591; "New England Electric Vehicle," *Electrical World* 36 (October 20, 1900): 662; "Automobiles," *Western Electrician* 28 (March 23, 1901): 214; "Failure of Electric Automobiles in Massachusetts," *Electrical World* 37 (April 6, 1901): 561–62; and "New England Electric Vehicle," *Electrical World* 42 (August 1, 1903): 223.
115. J. M. Hill to Walter H. Johnson, EVC, New York, November 24, 1899, Armstrong Papers, box 5, folder 7.
116. Report filed by Frank C. Armstrong, January 25, 1900; Armstrong to Walter H. Johnson, EVC, January 30, 1900; Report filed by A. W. Gilbert, "Estimates of Station Costs," February 10, 1900, Armstrong Papers, box 5, folder 7.
117. On the general history of Atlantic City, see Martin Paulsson, *The Social Anxieties of Progressive Reform: Atlantic City, 1854–1920* (New York: New York University Press, 1994); and Charles E. Funnell, *By the Beautiful Sea: The Rise and High Times of That Great American Resort, Atlantic City* (New York: Knopf, 1975).
118. Armstrong was eager to close the deals quickly, beseeching Johnson to come quickly: "no time is to be lost." Frank C. Armstrong to Walter H. Johnson, March 17, 1900, Armstrong Papers, box 6, folder 2.
119. See weekly revenue reports dating from July 7, 1900, Armstrong Papers, box 6, folder 2; and monthly balance statements, folder 5. Correspondence from a prospective customer queried at the end of the summer season underscored the damage done by the late start. Charles Hills wrote that "the station opened entirely too late to do any business" (Hills to New Jersey Electric Vehicle Transportation Company, New York, September 11, 1900, Armstrong Papers, box 5, folder 9).
120. See weekly revenue reports dating from July 7, 1900, Armstrong Papers, box 6, folder 2; and monthly balance statements, folder 5.
121. Woodward [first name unknown], Electric Storage Battery Co., to Walter H. Johnson, September 22, 1900, Armstrong Papers, box 5, folder 4.
122. C. C. Clark to Mr. Alden, EVC, September 28, 1900, Armstrong Papers, box 5, folder 4.
123. H. Crittenden to A. W. Gilbert, September 24, 1900, Armstrong Papers, box 5, folder 4.
124. Clark to Alden, EVC, September 28, 1900.
125. Walter H. Johnson to A. W. Gilbert, September 18, 1900, Armstrong Papers, box 5, folder 4.
126. Frank C. Armstrong to Walter H. Johnson, January 10, 1901, Armstrong Papers, box 6, folder 2.
127. "Automobiles for Hire," *Electrical World* 37 (February 16, 1901): 261. The vehicle counts probably overestimate the number of vehicles in service; lacking more accurate local reports, it is impossible to know the true extent of the specific fleets.
128. "Illinois Electric Vehicle Transportation Company to be Dissolved," *Western Electrician* 28 (March 9, 1901): 161; "Lead Cab Service Discontinued," *Horseless Age* 7 (March 6, 1901): 19; and "Electric Automobiles for Hire a Failure in Chicago," *Electrical World* 37 (March 16, 1901): 444.

129. "Automobiles in Chicago," *Electrical World* 37 (March 16, 1901): 429.
130. "Failure of Public Electric Automobiles in Massachusetts," *Electrical World* 37 (April 6, 1901): 561–62.
131. "The Status of Automobilism," *Electrical World* 37 (March 30, 1901): 501.
132. Comparative financial data offer a partial explanation for the decision not to continue service. During the best of times, return on capital was modest at the Atlantic City station, and the other branch stations along the Jersey shore (at Allenhurst, Long Branch, Seabright, and Spring Lake) hemorrhaged money from the start. At the end of February 1901, the New Jersey Electric Vehicle Transportation Company showed total assets of $93,031. With gross operating profits approaching $1,000 per month immediately before the closure of the New Jersey operation, the Atlantic City venture could have generated gross margins of 15 percent, a respectable if unspectacular performance.
133. See letter from Walter H. Johnson to Frank C. Armstrong, December 24, 1901, Armstrong Papers, box 3, folder 8; "Pennsylvania Electric Vehicle," *Electrical World* 38 (December 14, 1901): 1000; "New York Transportation," *Electrical World* 38 (December 28, 1901): 1082; and "New York Transportation" and "Pennsylvania Electric Vehicle," *Electrical World* 39 (March 8, 1902): 454.
134. Gijs Mom, "Inventing the Miracle Battery: Thomas Edison and the Electric Vehicle" (unpublished paper, 1998).
135. Palmer, "The Storage Battery," 647.
136. "Summary Expenses, Electric Cabs, 1903—1908," Meade Papers, box 26, folder "New York Transportation Company"; and "Public Passenger Service in New York," *Commercial Vehicle* 1 (June 1906): 124–29.
137. "Public Passenger Service in New York"; and memorandum from Richard W. Meade to board of directors, June 6, 1910, Meade Papers, Box 26, folder "New York Transportation."
138. Palmer, "The Storage Battery," 647.
139. "New York Transportation," *Electrical World* 41 (April 18, 1903): 672.
140. *Electrical World* 41 (April 18, 1903): 671; and "Biographical Note," finding aid, Meade Papers.
141. "New York Taximeter Cabs," *Horseless Age* 20 (September 4, 1907): 305; and Harry Perry, "Taximeter Cabs Actually in Use in New York," *Commercial Vehicle* 2 (July 1907): 174.
142. "Public Passenger Service in New York."
143. *Contract Automobile Livery Service*, n.d., Meade Papers, box 26, folder "New York Transportation." Records do not allow confirmation, but it is likely that the New York Transportation Company acquired the Newport station from the New England Electric Vehicle Transportation Company in 1901 or 1902. "Newport" first appears as a line item in the company budget in 1902: "Notebook of R. W. Meade—Operating Accounts and Other Statistics, 1900–1920," Meade Papers, box 15, folder "New York Transportation Company."
144. "Public Passenger Service in New York." 145. "New York Notes," *Western Electrician* 28 (June 15, 1901): 417; and "New York Transportation Association: A Short History," in official souvenir program, Meade Papers, box 15, folder "New York Transportation Company—7th Annual Employees Ball, 1909," 11.
146. The figures for overall profit and loss are more ambiguous due to unexplained miscellaneous entries that alter calculations of net profitability.
147. "Motor Cab Drivers' Strike," *Power Wagon* 2 (December 1906): 8–9.
148. On labor relations in the streetcar industry, see, for instance, Scott Molloy, *Trolley Wars: Streetcar Workers on the Line* (Washington, D.C.: Smithsonian Institution Press, 1996).
149. "Hundreds of Electric Cabs Burned," *Power Wagon* 3 (April 1907): 4.
150. "Taxicab Development in New York City," *Horseless Age* 21 (May 13, 1908): 571; "Taximeter Cab Situation in New York," *Commercial Vehicle* 2 (October 1907): 254–56; Perry, "Taximeter Cabs," 174–75; and "New York's Taximeter Cabs," *Horseless Age* 20 (September 4, 1907): 305.
151. *Power Wagon*, for instance, asked, "Why is it necessary to go abroad for motor vehicles for public service when it is highly probable that before full deliveries can be made American manufacturers will be freely producing machines in all respects as good as any which may be imported at twice the cost?" It went on to predict that "there are many disappointments in store for those who buy abroad before they can be brought to realize that their natural market is here at home." See "Motor Cabs for New York," *Power Wagon* 3 (May 1907): 5–6; and Joseph Anglada, "The Taxicab Business in New York," *Horseless Age* 24 (July 21, 1909): 63–65.
152. "Progress of the New York Taxicab Company," *Horseless Age* 20 (November 13, 1907): 737.
153. "Reducing the Repairs of Taxicabs," *Horseless Age* 21 (June 10, 1908): 699; and E. P. McDowell,

"Are Taxicab Companies Operating at a Profit?" *Horseless Age* 26 (July 27, 1910): 106. An editorial in *Horseless Age* claimed that average receipts for cabs fell by 50 percent as the "novelty wore off and the number of vehicles on the streets increased." See "Taxicab Development and Its Possible Consequences," *Horseless Age* 21 (May 13, 1908): 549.

154. "Causes of the Taxicab Strike," *Horseless Age* 22 (October 14, 1908): 517; and "The Motor Cab Drivers' Strike," 12–13.

155. The Sultan Taxicab Company claimed that its vehicle engines could be quickly exchanged with minimal effort. See "Sultan Taxicab with Removable Power Plant," *Commercial Vehicle* 5 (February 1910): 87–88. Also see the description of Bergdoll Motor Car Company taxis operating in Philadelphia in "The Taxicab Business in Philadelphia," *Horseless Age* 24 (August 11, 1909): 136.

156. Richard W. Meade, "Influence of Standardization on Taxicab Operation," *Horseless Age* 26 (July 27, 1910): 119–20. On the decision to suspend electric operations, see memorandum from chief engineer G. A. Green to Richard W. Meade, February 21, 1912, Meade Papers, box 15, folder "New York Transportation Company."

157. Meade wrote: "There is no possible doubt that, so far as accidents are concerned, the gasoline cab is a far better risk than the electric cab, and for this reason the amount of our accident and damage claims is decreasing as the electric cabs are eliminated. This is directly contrary to the belief of every accident insurance company representative with whom I have ever discussed the matter, as they believe that the higher speed of the gasoline cab makes it a more potent engine of destruction than the electric cab with its lower power and speed. The secret lies, however, in the lighter weight and better control of the gasoline can and the bad skidding propensities of the electric cab on solid tires" (memorandum from Richard W. Meade, n.d., Meade Papers, box 15, folder "New York Transportation Company"). An editorial in *Horseless Age* made a similar argument, claiming that the gasoline cab "would . . . be materially safer than the antiquated design of electric cabs now running ("Gasoline Cabs," 19 [March 6, 1907]: 371).

158. "Problems of the Automobile," *Electrical Engineer* 27 (February 2, 1899): 141.

159. On the competition between these interests, see, for instance, David J. St. Clair, *The Motorization of American Cities* (New York: Praeger, 1986); Stephen B. Goddard, *Getting There: The Epic Struggle between Road and Rail in the American Century* (Chicago: University of Chicago Press, 1994); and Scott Bottles, *Los Angeles and the Automobile: The Making of the Modern City* (Berkeley: University of California Press, 1987). For an alternative view, see Matt Roth, "Mulholland Highway and the Engineering Culture of Los Angeles in the 1920s" (paper presented at the annual meeting of the Society for the History of Technology, Pasadena, Calif., October 1997). The trade journals reported periodic conflict between horse-drawn cabs and the EVC. For instance, municipal law required that all vehicles that pick up passengers in the street display a numbered vehicle tag and a lamp. "As many of the customers of the automobile concerns object to riding in numbered cabs, the numbers have been removed, and as a result the Hack Owners Association has employed men to watch the electric vehicles" and report violations to the police. To avoid such problems, the cabs were "sent out only on special orders which come in through the offices." See "New York Notes," *Western Electrician* 27 (December 22, 1900): 403.

160. "Automobiles versus Trolley Cars," Edison National Historic Site, folder "Electric Storage Battery Company," reprinted courtesy of Gijs Mom. For sake of comparison, the EVC drivers were paid two dollars per day; the $2.50 fare was equal to fifty streetcar rides. A similar cab ride in 1995 might cost fifteen dollars, equal to ten subway or bus fares. The report also supports the claim that the rubber tire "always has been, is now and probably always will be the greatest element of expense and uncertainty. Whether the electric, steam, gasoline, compressed air, or other method of power is used, the expense of maintaining that part of the apparatus should always be the smallest as compared to the maintenance of the tire."

161. "Electromobiles and Electric Street Railways," *Electrical Review* 35 (August 16, 1899): 102.

162. "Electric Automobiles," *Electrical World* 32 (November 5, 1898): 465–66; and "The Electric Light Convention," *Electrical Review* 34 (May 31, 1899): 342.

163. C. E. Corrigan, "Condition of the Horseless Carriage Industry," *Western Electrician* 22 (January 1, 1898): 9.

164. An internal summary of the history of the company reported that "prior to July 1903 only fragmentary information is available, but this indicates a steady record of loss." In 1905 the electric buses were abandoned, and the company, unable to find a suitable internal combustion replacement, agreed to work with General Electric to develop a gasoline-electric motor bus. See "F.A.C.C. History," Meade Papers, box 17, folder "F.A.C.C."

165. "Statement of Average Buses Operated, Passengers Carried, Gross Operating Revenue," n.d., Meade Papers, box 17, folder "F.A.C.C." (emphasis in original).
166. "Revival of Electric Cab Service," *Power Wagon* 2 (August 1908): 5.

3. CENTRAL STATIONS AND THE ELECTRIC VEHICLE INDUSTRY

1. Neil Baldwin, *Edison: Inventing the Century* (New York: Hyperion, 1995), 138.
2. Prior studies of the relationship between the utilities and the electric vehicle industry are limited, although Sicilia (1990) includes a short section on the company's role in the electric vehicle campaign. Both Schiffer and Schallenberg treat the EVAA briefly. See David B. Sicilia, "Selling Power: Marketing and Monopoly at Boston Edison, 1886–1926" (Ph.D. diss., Brandeis University, 1990); Sicilia, "Selling Power: Marketing and Monopoly at Boston Edison, 1886–1926," *Business and Economic History* 20, 2d series (1991): 27–31; Richard Schallenberg, *Bottled Energy: Electrical Engineering and the Evolution of Chemical Energy Storage* (Philadelphia: American Philosophical Society, 1982), 280–83; and Michael Brian Schiffer, *Taking Charge: The Electric Automobile in America* (Washington, D.C.: Smithsonian Institution Press, 1994), 146–53.
3. R. McAllister Lloyd, "The Influence of the Pioneer Spirit on Electric Vehicle Progress," *Central Station* 14 (December 1914): 180. Schiffer, *Taking Charge,* includes a chapter titled "A Dark Age Descends," 91–102.
4. "Recent Developments in the Applications of Storage Batteries," *Electrical Review* 33 (August 17, 1898): 122; "The Growing Importance of the Storage Battery," *Electrical World* 32 (August 20, 1898): 178; and "The Storage Battery," *Electrical World* 35 (February 17, 1900): 241. Joseph Appleton reported industry-wide production of battery plates as follows:

Year	Weight of Plates (lbs.)
1894	349,000
1895	1,112,800
1896	3,315,300
1897	3,607,300

5. "A Powerful Storage Battery Combination," *Electrical Review* 25 (December 12, 1894): 298.
6. See the transcript of the discussion in *Electrical Review* 26 (February 27, 1895): 101–2. Later the same year, Charles L. Edgar, president of the Edison Illuminating Company of Boston, reported on the successful application of two large stationary batteries in Boston. See Charles L. Edgar, "Storage Battery Applications: Practical Experience with Storage Batteries in Central Stations," *Electrical Review* 27 (December 4, 1895): 316–17.
7. "It seldom occurs in a properly managed station that the period of minimum demand is long enough to allow of the laying off of one shift of men and the drawing of the fires. In most American stations, this period seldom amounts to more than four or five hours, so that . . . it is found more economical to charge the battery during this period than to shut down the engines and generators entirely." See Augustus Treadwell, Jr., "The Storage Battery," *Electrical World* 37 (January 5, 1901): 38–40.
8. Storage capacity was reported on an eight-hour discharge rate. See "The Chicago Edison Company's Storage Battery," *Electrical World* 31 (June 11, 1898): 726.
9. This was reprinted in abridged form as L. B. Stillwell, "Electrical Power Generating Stations and Transmission," *Electrical Review* 45 (October 29, 1904): 706.
10. "Electricity in 1904," *Electrical World* 46 (January 14, 1905): 48.
11. "Storage Batteries," *Electrical World* 55 (January 6, 1910): 8.
12. "Largest Central Station Storage Battery in the World," *Electrical World* 55 (February 11, 1909): 419; and "Changes in Storage Battery Practice," *Electrical World* 68 (August 8, 1916): 15. This change in practice was evident in how capacity was reported. Near the turn of the century, battery capacity was reported on the basis of an eight-hour discharge rate; by 1916 batteries were rated for one hour or, in the case of New York Edison, seven-minute discharge rates.
13. See the editorial "The Storage Battery," *Electrical World* 35 (February 17, 1900).
14. Harold L. Platt, *The Electric City: Energy and the Growth of the Chicago Area, 1880–1930* (Chicago: University of Chicago Press, 1991), 78. On Insull and the turbogenerator, see also Thomas P. Hughes, *Networks of Power: Electrification in Western Society, 1880–1930* (Baltimore: Johns Hopkins University Press, 1983), chap. 8. On the general growth of the electric industry during this period, see Richard F. Hirsh, *Technology and Transformation in the American Utility Industry* (New York: Cambridge University Press, 1989), 26–46; Harold C. Passer, *The Electrical Manufactures, 1875–1900* (Cambridge: Harvard University Press, 1953), chap. 21; William J. Hausman and John L. Neufield, "The Structure and Profitabil-

ity of the US Electric Utility Industry at the Turn of the Century," *Business History* 32 (April 1990): 225–43; and Paul A. David, "Heroes, Herds and Hysteresis in Technological History: Thomas Edison and 'The Battle of the Systems' Reconsidered," *Industrial and Corporate Change* 1, no. 1 (1992): 129–80. On load building and the turn to the electric, see Schiffer, *Taking Charge,* 99–107.

15. Platt, *Electric City,* chap. 4.
16. Hughes, *Networks of Power,* 209–14; and Platt, *Electric City,* 83–86.
17. Platt, *Electric City,* 102.
18. Ibid., 122.
19. Ibid., 151–61. See also David Nye, *Electrifying America: Social Meanings of a New Technology, 1880–1940* (Cambridge: MIT Press, 1990), chap. 6.
20. Platt, *Electric City,* 123.
21. Ibid., 137.
22. "Automobilism by Example," *Electrical World* 33 (March 11, 1899): 293.
23. "Charging Automobiles," *Electrical World* 36 (November 24, 1900): 795.
24. Elmer A. Sperry, "Automobiles As Source of Revenue for Central Stations," *Proceedings of the National Electric Light Association, Twenty-third Annual Convention* (Chicago, May 22–24, 1900), 373.
25. Thomas P. Hughes, *Elmer Sperry: Inventor and Engineer* (Baltimore: The Johns Hopkins University Press, 1971), 86–89.
26. A. E. Ridley, "The Electric Automobile As an Income Producer for Central Stations" (paper presented to the seventh annual convention of the Pacific Coast Electric Transmission Association), reprinted in *Central Station* 3 (October 1903): 85. Speaking to a western industry association, Ridley also pointed out that off-peak vehicle charging was especially valuable to central stations using power generated by free water.
27. "The Electric Automobile in Central Station Practice," *Central Station* 3 (April 1904): 244. See also L. D. Gibbs, "The Boston Edison Company and the Electric Vehicle Business," *Electrical World* 53 (February 18, 1909): 453–54.
28. Henry Cave was the engineer in charge of building the vehicle and described driving the truck from the Harlem railyards to the Edison offices with the brakes locked for storage. See Henry Cave to A. N. Dingee, Electric Storage Battery Company, February 21, 1945, Henry Cave Papers, National Automotive History Collection, Detroit Public Library, box 3, folder 2.
29. *Horseless Age* 10 (August 20, 1902): 204.
30. "Relative Merits of Gasoline and Electric Vehicles," *Motor Traffic,* April 16, 1906, Hiram Percy Maxim Papers, Connecticut State Library, Hartford, box 5, pp. 9–10.
31. An *Electrical World* editorial in 1905 doubted "whether 25% of the central station managers in this country have ever yet taken enough interest in electric automobiles to get one, either for pleasure or for business. . . . They would be encouraging their own industry, learning something about the subject, and setting a good example to many possible customers." See "Electric Automobiles," *Electrical World* 46 (August 19, 1905): 293–94.
32. The following journals were started: *Electric City* (Chicago, 1903), *Bulletin of the New York Edison Company* (New York, 1902), *Edison Light* (Boston, 1903), *Brooklyn Edison* (Brooklyn, 1903), and *Current News* (Philadelphia, n.d.).
33. See photographs in "The Rapidly Increasing Use of Electric Vehicles in Connection with Central Station Practice," *Central Station* 7 (November 1907): 734, 735; and "Edison Delivery Service," *Central Station* 8 (March 1909): 213.
34. "Electrical Vehicle Manufacturers," *Horseless Age* 18 (December 12, 1906): 878.
35. Hayden Eames, "The Electric Wagon," *Power Wagon* 2 (April 1907): 11–12.
36. Herbert H. Rice, "Opportunities for the Sale of Current for Charging Electric Automobiles," *Proceedings of the National Electric Light Association, Thirtieth Annual Convention* (Indianapolis, June 6, 1907), 498.
37. Ibid., 507.
38. Frank J. Stone, "The Electric Vehicle Situation in New England," *Electrical Review* 56 (February 19, 1910): 391–92.
39. For the motto, see any of the meeting reports in *Central Station* 9 and 10. Once the Boston organization gave way to the national Electric Vehicle Association of America, the institutional agenda expanded and the motto was broadened: "To promote the Adoption and Use of Electric Vehicles for Business and Pleasure Purposes." See *Central Station* 10 (October 1910): 98.
40. "Chicago Automobile Show," *Electrical Review* 54 (February 13, 1909): 287.

41. "Boosting the Use of Electric Vehicles in New England," *Electrical Review* 54 (March 6, 1909): 422.
42. "Chicago Automobile Show," *Electrical Review* 54 (February 13, 1909): 287.
43. "The December Meeting," *Central Station* 9 (January 1910): 142–44; and "The Electric Storage Battery Company Dinner," *Central Station* 9 (February 1910): 172–73.
44. "The Electric Vehicle Association of America," *Central Station* 9 (June 1910): 293–94; "The Electric Vehicle Association of America," *Central Station* 10 (July 1910): 18–19; and "Electric Vehicle Association of America," *Central Station* 10 (September 1910): 79–80.
45. E. S. Mansfield, "The Electric Vehicle Proposition," *Central Station* 11 (August 1911): 43.
46. E. S. Mansfield, "The Central Station Back of the Electric Vehicle," *Proceedings of the Second Annual Convention of the Electric Vehicle Association of America* (New York City, October 10, 1911), 28–29. The Atlantic Avenue garage was ostensibly operated under the auspices of the EVAA, but there is no indication that the association was actively involved in its day-to-day activities. Instead, it seems to have been the first of several city garages operated with the approval of the national body. See "The Boston Edison Campaign," *Central Station* 10 (April 1911): 285; and Sicilia, "Selling Power" (1990), 345–50. On Edgar's personal vehicle, see "The Electric Vehicle and the Central Station," *Electrical World* 54 (December 2, 1909): 1339.
47. William C. Anderson, "How a Central Station Can Develop Its Electric Vehicle Load," *Central Station* 13 (July 1913): 24–25.
48. "The October Meeting, " *Central Station* 9 (November 1909): 94; and "The New York Edison Company and the Electric Vehicle," *Central Station* 9 (March 1910): 197.
49. "Address of Mr. Arthur Williams, President of the Electric Vehicle Association of America, read by Mr. Hendershot at the Annual Meeting of the Chicago Section of the Electric Vehicle Association of America," *Central Station* 13 (July 1913): 21.
50. James T. Hutchings, "Attitude of Central Stations toward Electric Automobiles," *Proceedings of the First Annual Convention of the New England Section, National Electric Light Association* (Boston, March 16, 1910), 26–27.
51. "Electric Automobiles at Denver," *Electrical World* 58 (August 26, 1911): 498–500; and "Electric Vehicles Association of America, Chicago Section," *Central Station* 12 (April 1913): 341–42. See also Ross B. Mateer commenting on "Report of Committee on Electric Vehicles," *Proceedings of the National Electric Light Association, Thirty-fifth Annual Convention* (Seattle, June 12, 1912), 251–52. Here and in similar instances, page numbers refer to transcripts of discussions, not the papers themselves.
52. C. E. Michel, "History of the Electric Vehicle Business in St. Louis," *Central Station* 9 (April 1910): 218–20; comments, *Central Station* 12 (February 1913): 268.
53. "The Relations between Central Stations and the Electric Automobile," *Electrical World* 54 (August 5, 1909): 314–20.
54. "The Electric Vehicle Campaign of the Edison Electric Illuminating Company of Boston," *Central Station* 10 (May 1911): 320.
55. "Report of Publicity Committee on the National Co-operative Advertising Campaign of the Electric Vehicle Association of America," *Proceedings of the Third Annual Convention of the Electric Vehicle Association of America* (Boston, October 9, 1912), 13.
56. Herbert H. Rice, discussing a paper by T. Commerford Martin and Kingsley Martin, "Has the Electric Fully Arrived?" *Central Station* 11 (December 1911): 178.
57. P. D. Wagoner, discussing a paper by William P. Kennedy, "Administrative Engineering and Salesmanship in the Commercial Car Field," *Central Station* 11 (July 1911): 23.
58. M. Holland, discussing a paper by William P. Kennedy, "Administrative Engineering and Salesmanship in the Commercial Car Field," 22.
59. Willis M. Thayer, "The Hartford Electric Light Company's Experience with the Battery Exchange System for Commercial Vehicles," *Central Station* 15 (December 1915): 141; and William Blood, discussing a paper by T. Commerford Martin and Kingsley Martin, "Has the Electric Fully Arrived?" 176.
60. William P. Kennedy, responding to comments on "Administrative Engineering and Salesmanship in the Commercial Car Field," 23–24.
61. M. Robertson, discussing a paper by H. W. Hillman, "Relative Importance of the Electric Truck As Compared with Other Classes of Central Station Business," *Central Station* 12 (February 1913): 260.
62. M. Shoepf, discussing a paper by Alexander Churchward, "The Standardization of the Electric Vehicle, Part II (Speed)," *Central Station* 11 (March 1912): 260–61.
63. Ellis L. Howland, "Practical Ideals in Electric Vehicle Promotion," *Central Station* 14 (October 1914): 100.

64. Repeat orders and the distribution of installations were described as follows: "Less than fifty of the larger electric installations include over 10% of all the commercial power vehicles in operation in the country, which 10% assumes gigantic proportions when it is considered the other 90% are distributed among nearly 7,000 users." See "The Electric Commercial Vehicle," *Central Station* 12 (July 1912): 30. Also note that the Baker Company claimed that 70 percent of its business between mid-1911 and mid-1912 consisted of reorders. See "Baker Electric Trucks," *Central Station* 12 (August 1912): 57.

65. "Training Dealers to Sell Electric Vehicles," *Electrical World* 65 (February 27, 1915): 553.

66. "The December Meeting of the Electric Vehicle Association of America," *Central Station* 10 (January 1911): 191.

67. James K. Pumpelly, "Standardization of Automobile Batteries," *Western Electrician* 26 (April 7, 1900): 221.

68. "Automobile Standardization," *Electrical World* 35 (February 24, 1900): 275.

69. "Charging Automobiles," *Electrical World* 36 (November 24, 1900): 795.

70. Day Baker, "Desirability of a Standard Charging Plug for All Electric Vehicles," *Central Station* 10 (September 1910): 77.

71. Ibid.

72. Following the third EVAA annual meeting, some minor issues were still unresolved, and the final designs were not announced until early 1914. See "E.V.A. Standardizes Charging Plugs," *Commercial Vehicle* 10 (February 1, 1914): 9.

73. "June Meeting of the Electric Vehicle Association of America," *Central Station* 14 (July 1914): 17. In "More 'Small Things Forgotten': Domestic Electrical Plugs and Receptacles, 1881–1931" (*Technology and Culture* 27 [July 1986]: 525–48), Fred E. H. Schroeder examines the emergence of the standard household outlet. It is interesting to note that the EVAA plug preceded the regular commercial standard by five years and that, while the main commercial producers had to wait for the pressures of World War I to drive them to standardize, the EVAA was able to independently orchestrate acceptance of a standard.

74. Schroeder's article (see note 73) focuses almost exclusively on the emergence of compatibility and misses other interesting aspects of the process of electrical standardization.

75. "New England Section, Electric Vehicle Association of America," *Central Station* 13 (July 1913): 23.

76. Robert M. Lloyd, "Standardization of Commercial Vehicles," *Central Station* 11 (March 1912): 264.

77. Consulting engineer William P. Kennedy (still involved with the electric vehicle trade after advising on pioneering ventures such as the Electric Vehicle Company, Studebaker Brothers, and the Baker Motor Car Company) was both an EVAA director and a member of the Society of Automobile Engineers committee of vehicle standardization, where standards for tires, lighting, and batteries were all discussed. See "The Electric Vehicle Association of America," *Central Station* 14 (February 1915): 243; "Centralizing Batteries for Electric Trucks," *Central Station* 15 (September 1915): 65–66; and William P. Kennedy, "Vehicle Standardization Possibilities," *Commercial Vehicle* 6 (February 1911): 143.

78. A. Jackson Marshall, "Annual Report of the Secretary of the Electric Vehicle Association of America for the Fiscal Year Ended Sept. 30, 1915" (paper presented at the sixth annual convention of the Electric Vehicle Association of America, October 18–19, 1915), 19–20.

79. Churchward, "The Standardization of the Electric Vehicle," 254.

80. Sperry, "Automobiles As Source of Revenue for Central Stations," 374.

81. Virginia Scharff, *Taking the Wheel: Women and the Coming of the Motor Age* (New York: Free Press, 1991): 46.

82. William P. Kennedy, discussing a paper by G. H. Jones and R. Macrae, "The Electric Passenger Vehicle," *Central Station* 11 (June 1912): 360.

83. Schiffer, *Taking Charge*, 153–55. The research center at the Henry Ford Museum and Greenfield Village has a thick sheaf of letters from people interested in either buying or selling the rumored Ford electric (see chapter 6).

84. J. W. Brown, comments, *Central Station* 12 (September 1912): 84.

85. William P. Kennedy, discussing "Report of Committee on Operating Records," *Proceedings of the Fourth Annual Convention of the Electric Vehicle Association of America* (Chicago, October 27, 1913), 9.

86. "The Fifth Annual Convention of the Electric Vehicle Association of America," *Central Station* 14 (November 1914): 158.

87. "Users Get Together, *Commercial Vehicle* 14 (May 1, 1916): 12.

88. Kennedy, discussing "Report of Committee on Operating Records," 9.

89. "Operating Costs of Electric Vehicles" and "Operating Costs for Commercial Electric Vehicles," *Electrical World* 64 (October 3, 1914): 645; 664–65.

90. Part of this fundamental uncertainty resulted from the difference between organic (horse), mechanical (internal combustion), and electrochemical (battery) systems. The state of the first two could be determined by more or less direct observation, whereas indirect instrumentation and subsequent interpretation was necessary to determine the state of charge of a storage battery. Hydrometers, for instance, reported on the state of charge of a single cell; but unless the operator was willing to take roughly forty to sixty measurements, there was no way to know the overall condition of the battery. Similarly, the "foolproof" Sangamo meter was intended to ease the process of measuring current entering and leaving the battery; but it, too, was often unreliable. As one EVAA member revealed at a monthly meeting in 1913, a company depending on the results from thirty-five Sangamo meters subsequently discovered that thirty-four of them had been improperly connected and were therefore producing unreliable readings. See R. C. Lanphier, "The Ampere-Hour Meter for Electric Vehicles," *Central Station* 10 (May 1911): 311–19; "Sangamo Meters," *Central Station* 12 (August 1912): 57; and M. Martin, discussing a paper by James M. Skinner, "The Philadelphia Thin Plate Battery," *Central Station* 12 (March 1913): 306.

91. On the idea of the heterogeneous engineer, see John Law, "Technology and Heterogeneous Engineering: The Case of Portuguese Expansion," in *The Social Construction of Technological Systems*, edited by Wiebe E. Bijker, Thomas P. Hughes, and Trevor Pinch (Cambridge: MIT Press, 1987), 111–34.

92. In this connection, see William P. Kennedy, discussing a paper by Harold Pender and H. F. Thomson, "Observations on Horse and Motor Trucking": "About five years ago [1908] those of us here who are interested . . . had great difficulty in finding a concrete or reliable record of those horse equipments' operating cost, because the owners of these equipments usually neglected the stable. They thought the stable equipment was rather beneath their dignity and never gave any particular attention to that department of their institution. . . . we have awakened them to the realization that they have spent a great deal more money in the operation of horse equipment than they previously realized" (*Central Station* 12 [April 1912]: 333).

93. Louis A. Ferguson, "The Responsibility of the Central Station to the Electric Vehicle Industry," *Proceedings of the First Annual Convention of the Electric Vehicle Association of America* (New York, October 18, 1910), reprinted in *Central Station* 10 (November 1910): 142.

94. William P. Kennedy, "Problems Involved in Advancing the Use of Electric Vehicles," *Proceedings of the First Annual Convention of the Electric Vehicle Association of America*, 138–40 (emphasis in original).

95. Ferguson, "The Responsibility of the Central Station," 142. Herbert Rice was in the audience for Ferguson's talk and asked the EVAA's President Blood to promise that the proceedings would be published and widely disseminated. Sitting in the audience at the first EVAA convention, Rice must have taken some solace in hearing the central station representatives own up to their past negligence.

96. See my previous discussion of Lloyd's proposed monopoly. Day Baker, president of the Boston Electric Motor Car Club, proposed a similar plan in which the club would help establish an "electric motor trucking express and parcel delivery company" with paid-in capital of $100,000. Boston Edison would promote the company but not invest in it. See *Central Station* 12 (July 1912): 26.

97. "Report of Committee on Electric Vehicles," *Proceedings of the National Electric Light Association, 34th Annual Convention* (New York, June 1, 1911), 593.

98. William C. Anderson counseled central stations: "The establishment of a responsible dealer who is capable and has the proper facilities for handling the electric vehicle business is the FIRST STEP towards success. If such a dealer cannot be easily located in your city, you should take hold of the proposition yourself until an excellent representative of the proper calibre could be found" ("How a Central Station Can Develop Its Electric Vehicle Load," 24). According to the "Report of the Committee on Electric Vehicles," as of 1912, only seven out of the forty-nine central stations using electric vehicles were agents for specific vehicle lines. See *Proceedings of the National Electric Light Association, Thirty-fifth Annual Convention*, 247.

99. Mansfield, "The Electric Vehicle Proposition," 43.

100. R. Macrae, "Ideal Electric Garage Service," *Central Station* 12 (May 1913): 370.

101. Stephen G. Thompson, discussing another paper on the ideal electric garage, described his outrage after receiving a thirty-three-dollar bill for service that was necessary "due to negligence on the part of the garage." See "May Meeting of the Electric Vehicle Association," *Central Station* 12 (June 1913): 408.

102. J. C. Bartlett, discussing a paper by William Elwell, "Care of Electric Vehicles in the Garage of the Philadelphia Electric Company," *Central Station* 13 (May 1914): 430.
103. Carroll A. Haines, "I Should Worry!" *Central Station* 13 (May 1914): 432; and Haines, "From the Garage Owner's Point of View: A Few of the Problems Which Confront Garagemen in His Attempt to Give Service Satisfactory to All," *Electric Vehicles* 4 (May 1914): 199.
104. Stephen McIntyre, "'The Repair Man Will Gyp You': Mechanics, Managers and Customers in the Automobile Repair Industry" (Ph.D. diss., University of Missouri–Columbia, 1995).
105. "Electric Vehicle Association of America," *Central Station* 15 (October 1915): 99.
106. Macrae, "Ideal Electric Garage Service," 371: "For years the main talking point of the electric vehicle salesman has been that it is not necessary to have an experienced man to operate an electric vehicle."
107. See the discussion of a paper by Daniel J. Tobin, president of the International Brotherhood of Teamsters, "April Meeting of the Electric Vehicle Association of America," *Central Station* 12 (May 1913): 360–67.
108. "Speed Maniac New Menace to Trucks," *Central Station* 11 (June 1912): 367–68. New Jersey Public Service Company executive Stephen G. Thompson also supported this idea: "Because of the characteristics of the motor, which tends to slow down under excessive current flow, the electric machine is not subject to the effect of the abuses of an unskilled operator" ("The Electric Vehicle," *Central Station* 13 [August 1913]: 48).
109. Macrae, "Ideal Electric Garage Service," 371; and Mansfield, "The Electric Vehicle Proposition," 45.
110. Lack of enrollment is perhaps explained by the fact that the course announcement did not mention who was to pay the ten-dollar tuition fee. Without help from either Chicago Edison or the Electric Vehicle Association, drivers and garage employees were not likely to pay more than half a week's wages to enroll in the class. See "Electric Vehicle Course at Armour Institute," *Electrical World* 64 (September 19, 1914): 552;
111. "Electric Vehicle Technical School Started in Chicago," *Electrical World* 75 (June 12, 1920): 1379; and H. E. Sandoval, "Selling Trucks on the Pacific Coast," *Proceedings of the National Electric Light Association, Forty-seventh Annual Convention* (Atlantic City, May 22, 1924), 619. Forty-two students enrolled in the San Francisco course.
112. E. J. Bartlett, "Utility of Passenger and Commercial Electrics," *Electric Vehicles* 5 (July 1914): 3–5.
113. "January Meetings of the Electric Vehicle Association of America," *Central Station* 11 (February 1912): 227.
114. William Blood, discussing a paper by Jones and Macrae, "The Electric Passenger Vehicle," 359.
115. "A Frank Statement," *Electric Vehicles* 4 (May 1914): 177–78.
116. John F. Gilchrist, "Opportunities for the Central Station in the Electric Vehicle Industry in 1915," *Electrical World* 65 (January 2, 1915): 9–10.
117. "The Electric Vehicle Convention," *Central Station* 15 (November 1915): 129.
118. "E.V.A.A. Absorbed by the N.E.L.A.," *Central Station* 15 (March 1916): 233.
119. U.S. Bureau of the Census, *Abstract of Census of Manufactures, 1914* (Washington, D.C.: General Printing Office, 1917), 226.
120. William P. Kennedy, "Central Station Promotion of Electric Vehicle Use," *Central Station* 15 (June 1916): 321–24.
121. Ibid., 322.
122. "The Fifth Annual Convention of the Electric Vehicle Association of America," *Central Station* 14 (November 1914): 145; "The Central Station and the Electric Vehicle," *Central Station* 14 (March 1915): 261; and A. Jackson Marshall, "The Electric Vehicle—1916–1917," *Central Station* 16 (January 1917): 173.
123. Frank Smith, cited in "Small Stations and the Electric Vehicle," *National Electric Light Association Bulletin*, new series 1 (December 1914): 701. For "gradual" and "steady," see A. Jackson Marshall, "Review of Electric Vehicle Industry and Forecasts for the Coming Year," *Central Station* 15 (January 1916): 170.
124. Thompson, "The Electric Vehicle," 47.
125. "The Evolution of the Electric Truck an Important Factor in Delivery Problems," *Central Station* 10 (May 1911): 323.
126. Arthur Williams, "The Development of the Electric Vehicle," *Central Station* 12 (February 1913): 233.
127. "Why the Electric Succeeds," *Central Station* 11 (March 1912): 262. This line of reasoning is examined in detail in chapter 4.

128. Anderson, "How a Central Station Can Develop Its Electric Vehicle Load," 24. By way of comparison, leading stations were offering charging current for five cents or less per kilowatt-hour.
129. "Address of Mr. Arthur Williams," 19.
130. M. C. Duffy, "The Mercury-Arc Rectifier and Supply to Electric Railways," *Engineering Science and Education Journal* 4 (August 1995): 183–92.
131. "Baker and R&L One," *Commercial Vehicle* 12 (June 15, 1915): 16; and R. Thomas Wilson, *First Hundred Years. Baker Raulang* (Cleveland: Baker Raulang, 1953), in the collection of the Western Reserve Historical Society, Cleveland.
132. T. Commerford Martin, quoted in *Electrical World* 53 (June 10, 1909): 1483.
133. Sicilia, "Selling Power" (1990), 342.
134. "The November Meeting," *Central Station* 9 (December 1909): 117 (emphasis in original).
135. C. W. Squires, Jr., "The Future of the Electric Truck," *Central Station* 15 (December 1915): 158.
136. "The Greater Electric Vehicle Problems," *Electric Vehicles* (July 1914): 25.

4.   THE ELECTRIC TRUCK AND THE RISE AND FALL OF
     APPROPRIATE SPHERES

1. "Why Not a Real Demonstration?" *Commercial Vehicle* 12 (February 1, 1915): 10.
2. Bruce E. Seely, *Building the American Highway System: Engineers As Policy Makers* (Philadelphia: Temple University Press, 1987), chap. 3; Wilfred Owen, *The Metropolitan Transportation Problem* (Washington, D.C.: Brookings Institution, 1956), chap. 2; Mark Rose, *Interstate: Express Highway Politics, 1939–1990* (Knoxville: University of Tennessee Press, 1990 [1979]); Robert Raburn, "Motor Freight and Urban Morphogenesis With Reference to California and the West" (Ph.D. diss., University of California, Berkeley, 1988); Spencer Miller, "History of the Modern Highway in the United States," *Highways in Our National Life*, ed. Jean Labatut and Wheaton J. Lane (Princeton, N.J.: Princeton University Press, 1950), 88–119; and George Romney, "The Motor Vehicle and the Highway: Some Historical Implications," also in *Highways in Our National Life*, 215–26.
3. F. M. L. Thompson reports that in 1893 the railroads employed 6,000 horses in London alone. See his *Horses in European Economic History: A Preliminary Canter* (London: British Agricultural Society, 1983), 102.
4. "Parcel Delivery Discussed by Truck Club," *Commercial Vehicle* 9 (October 1, 1913): 13.
5. In Cleveland, for instance, Alexander Winton's first vehicles had been used in his own factory; and by the late 1890s progressive merchants in a variety of industries had begun to replace some of their horse-drawn wagons with experimental gasoline, steam, and electric vehicles. Of course, the electric taxicabs and omnibuses made by the Electric Vehicle Company had also provided commercial service. On the whole, however, the earliest motor vehicles were sold to individuals for pleasure driving. A number of professionals, especially doctors, used their cars for business purposes, but the transport service they demanded differed from pleasure driving only in quantity, not quality.
6. "The Commercial Vehicle," *Commercial Vehicle* 1 (March 1906): 28.
7. Editorial excerpted in "The Horseless Town," *Power Wagon* 2 (April 1907): 22.
8. See pictures in John Gilchrist and A. Jackson Marshall, "The Electric Vehicle and the Central Station," *Proceedings of the National Electric Light Association, Thirty-eighth Annual Convention* (San Francisco, June 7–11, 1915), 337.
9. *Electrical World* claimed that B. Altman was the first company in New York to use automobile delivery wagons; reportedly, the company put its first vehicle in service in early 1898 (see "Delivery Automobile," *Electrical World* 35 [May 26, 1900]: 806; and Max Lowenthal, "Electric Delivery Equipment of a Large New York Store," *Electrical World* 36 [December 29, 1900]: 987–89). The first vehicles purchased by Gorham in 1899 were made by Andrew Riker (see Howard Greene, "An Object Lesson in Electric Delivery," *Commercial Vehicle* 1 [March 1906]: 90; and "Gorham Electric Garage Is Truck Factory," *Commercial Vehicle* 9 [October 15, 1913]: 38). For Abraham and Straus, see "Machines Effect Saving in Dry Goods Delivery," *Commercial Vehicle* 6 (June 1911): 301–4. Also see reports on length of service in the survey article, "Electric Commercial Trucks Are Increasing Throughout the Country," *Commercial Vehicle* 9 (October 15, 1913): 5–11.
10. "Operating the Electric Vehicle; Cost Figures, Trips, Maintenance," *Commercial Vehicle* 7 (November 1912): 9.
11. So-called "transfer" service also employed large-capacity vehicles. Chicago's Montgomery

Ward, for instance, had been operating six five-ton electric trucks in transfer service since 1903. "Gasoline and Electric Trucks in Chicago," *Commercial Vehicle* 1 (March 1906): 31.

12. "How Chicago's Coal Supply Is Handled," *Commercial Vehicle* 7 (March 1912): 59.

13. "Present State of the Commercial Automobile Movement Throughout the Country," *Horseless Age* 16 (July 5, 1905): 9.

14. Henri G. Chatain, "The Liverpool Trials," *Horseless Age* 8 (July 10, 1901): 327.

15. Robert Raburn has examined several early attempts to use steam vehicles for ore and lumber transport. None of the applications succeeded, however, leading him to conclude that steam trucks (or tractors) offered "geographically imperfect" service. A number of short-haul steam wagons were tried before 1903, but none remained in use. See Raburn, "Motor Freight and Urban Morphogenesis," 66.

16. Raburn argues that support for separate spheres for the different motive powers resulted from the marketing and promotional efforts of electric vehicle industry groups such as NELA and the EVAA. In particular, he discounts the findings of the group of MIT-based transport engineers because their research was funded by Boston Edison. The MIT studies will be examined in detail later in this chapter, but for the moment it is important to see that this analysis offers a different explanation for the rise of appropriate spheres. Belief in functional specialization predated by several years the organized activities of NELA and the EVAA in support of electric vehicles. Clough's article appeared in 1903, while the Electric Vehicle and Central Station Association was not organized until 1909.

17. Thomas P. Hughes, *American Genesis* (New York: Penguin, 1989), 188–203.

18. "Operating the Electric Vehicle," 5–11.

19. "Scorning Motion Study," *Commercial Vehicle* 8 (May 1, 1913): 13. Literally dozens of studies were performed that purported to have developed better and more thorough oversight systems. An article on the Gimbel Brothers department store warned, "The figures without the system are bewildering, and *the system is the starting point*" (see "Keeping Track of Cost of Operation," in the section "Systems for Maker and User," *Commercial Vehicle* 7 [December 1912]: 21; emphasis in original).

20. "Short-Haul Problems," *Commercial Vehicle* 10 (June 15, 1914): 12. By reorganizing transportation service, merchants were following the journal's call to "sciencize" their operations ("Sciencize Truck Operation," *Commercial Vehicle* 12 [April 1, 1913]: 12).

21. "What Becomes of the Horses and Wagons? With Special Reference to the Loss, If Any, Which a Merchant Must Stand in Adopting Motor Trucks," *Commercial Vehicle* 7 (February 1912): 38–39.

22. Editorial excerpted in "The Horseless Town," 22.

23. "What Motor Trucks Are Actually Doing for American Users," *Commercial Vehicle* 10 (February 15, 1914): 5–14.

24. Joseph Husson noted that this "custom" was "firmly rooted in all the truck drivers and was handed down from the horse-wagon days." See "Brewery Trucks Reduce Delivery Cost and Increase Business Area, Part I," *Commercial Vehicle* 11 (November 1, 1914): 6.

25. Ibid.

26. "Choosing the Type," *Electric Vehicles* 4 (March 1914): 96.

27. Clinton Woods, "Future of the Self-Propelled Vehicle," *Western Electrician* 21 (December 4, 1897): 315–17.

28. "Transportation by Machine," *Commercial Vehicle* 2 (November 1907): 276–77.

29. "Business Methods for Business Vehicles," *Commercial Vehicle* 8 (May 1, 1913): 29.

30. A. N. Bingham, "Operating Costs of Gas Motor Trucks," *Commercial Vehicle* 6 (August 1911): 415.

31. "Why Not a Real Demonstration?" 11.

32. An article titled "Selling the Delivery" noted, "The purchaser of two or three delivery motor vehicles has many difficulties that confront him; it is the nightmare of these difficulties that is holding him back from purchasing" (*Commercial Vehicle* 14 [March 1, 1916]: 12).

33. William B. Stout, "Philosophy of the Light Delivery Car—Its Field and Future," *Commercial Vehicle* 12 (February 1, 1915): 9.

34. "Simpler, More Flexible and Lighter Trucks in 1915 Offerings," and "The Trends of 1915," both in *Commercial Vehicle* 11 (January 1, 1915): 7 and 42–43.

35. "Great Production Prevents Radical Changes for 1917," *Commercial Vehicle* 15 (November 1, 1916): 13.

36. Albert Clough, "General Review of the Commercial Vehicle Problem," *Horseless Age* 16 (July 5, 1905): 4.

37. E. J. Kilbourne, "The Relation of the Driver to the Vehicle," *Electric Vehicles* 4 (April 1914): 139.

38. "The Trend of the Times in Truck Work," *Commercial Vehicle* 7 (March 1912): 25–26.

39. G. Hives Dawson, "Steam As a Motive Power for Trucks," *Commercial Vehicle* 1 (May 1906): 94.

40. E. J. Bartlett, "Utility of Passenger and Commercial Electric Vehicles," *Electric Vehicles* 5 (July 1914): 3–5.

41. E. J. Bartlett, "When and Why Electric Trucks Are Economical," *Electric Vehicles* 4 (January 1914): 5–6.

42. Although electric trucks were less susceptible to abuse from speeding, drivers could still damage vehicle and cargo. To monitor driver behavior, some owners fitted their vehicles with a coastometer—a device originally developed for street railways that measured the distance traveled without power (that is, coasting) as a fraction of total mileage. A driver who was careful with current could coast more than one who was wasteful (see "Coastometer Detects Driver's Carelessness," *Commercial Vehicle* 10 [February 1, 1914]: 42; "Coasting Ability of Electrics," *Electric Vehicles* 4 [May 1914]: 163; and "The Advantages of the Coastometer," *Electric Vehicles* 5 [August 1914]: 55). The speedograph (another monitoring technology) was also tried and in at least one case identified a brewery truck driver who "had been loafing and consuming the brewery's product for approximately thirty minutes at each stop and had been exceeding the speed limit and pushing the truck to its utmost capacity between stops with absolutely no regard to road conditions or other factors which should influence a driver's judgment" (see "Remedy for Electric Vehicle Trouble," *Electrical World* 62 [September 6, 1913]: 484).

43. "Training Dealers to Sell Vehicles," *Electrical World* 65 (February 27, 1915): 553.

44. William B. Stout, "Motor Trucks in Packing Industry," *Commercial Vehicle* 7 (May 1912): 8.

45. "The Duplex Governor," *Commercial Vehicle* 10 (February 15, 1914): 31.

46. "Progress and Trends of One Year in Commercial Vehicle Industry," *Commercial Vehicle* 9 (January 1, 1914): 7. An editorial on the subject in 1914 criticized the industry for its "lukewarm" efforts and called for the development of a "genuine governor, not a governor that can be adjusted by the driver or owner" (see "Truck Trends to Date," *Commercial Vehicle* 9 [January 1, 1914]: 45).

47. "Truck Speedometer Registers Maximum Speed," *Commercial Vehicle* 10 (February 1, 1914): 42.

48. "Careless Drivers," *Power Wagon* 2 (May 1907): 7.

49. "Plans to Reduce Motor Truck Abuse," *Electric Vehicles* 5 (July 1914): 22.

50. "Why the Driver Problem Is Difficult," *Commercial Vehicle* 8 (May 15, 1913): 12; and "A Square Deal: Basis of Successful Bonus System for Drivers," *Commercial Vehicle* 15 (October 15, 1915): 16–19.

51. "Do Not Make Civil War," *Commercial Vehicle* 7 (November 1912): 36.

52. "The Motor Truck Industry As It Is: Makers, Models, Users and Sales," *Commercial Vehicle* 8 (January 1913): 5.

53. W. L. Day of the General Motors Truck Company quoted in "National Automobile Show," *Electric Vehicles* 4 (January 1914): 4.

54. The purchase of the Landsen company prompted an editorial envisioning an "ideal" market in which "all of the best firms would build both kinds in all sizes so as to have a complete line and thus compete for any order" (see "The Two Kinds of Truck Power," *Commercial Vehicle* 7 [March 1912]: 49).

55. "Features of GMC Electric Trucks," *Commercial Vehicle* 7 (May 1912): 42–43. General Motors electric trucks ceased to appear in model year 1917 ("A Directory of Motor Truck Makers," *Commercial Vehicle* 15 [November 1, 1916]: 17–43). A recent article by GM engineer Kaushik Rajashekara inexplicably claims that GMC Truck *began* producing electric trucks in 1916. See Rajashekara's "History of Electric Vehicles in General Motors," *IEEE Transactions on Industry Applications* 30 (July–August 1994): 897.

56. "The Simple Truth," *Commercial Vehicle* 10 (May 1, 1914): 14.

57. Merrill C. Horine, "The Motor Truck and Why," *Commercial Vehicle* 12 (July 15, 1915): 8.

58. "The Fifth Annual Convention of the Electric Vehicle Association of America," *Central Station* 14 (November 1914): 154.

59. "The Central Station Opportunity in Electric Vehicles," *NELA Bulletin*, n.s. 1 (July 1914): 441.

60. Walter C. Reid, "Electricity As a Substitute for Horses in Local Removals," *Central Station* 12 (August 1912): 52.

61. Harold Pender and H. F. Thomson, "Observations on Horse and Motor Trucking," *Central Station* 12 (April 1913): 331.

62. "Federal Government Buys Electric Trucks," *Central Station* 12 (August 1912): 58.
63. Day Baker, "Electric Vehicle Fleets in the United States," *Commercial Vehicle* 6 (May 1911): 252.
64. William Blood, discussing "Report of the Committee on Electric Vehicles," *Proceedings of the National Electric Light Association, Thirty-fifth Annual Convention* (Seattle, June 12, 1912), 255.
65. Pender and Thomson, "Observations on Horse and Motor Trucking," 330.
66. Edward E. La Schum, *The Electric Motor Truck: Selection of Motor Vehicle Equipment, Its Operation and Maintenance* (New York: U.P.C. Book Company, 1924), 12, 285.
67. "Efficient Work of Commercial Vehicles," *Commercial Vehicle* 8 (May 15, 1913): 22.
68. "What the Ton-mile Is and How to Figure It," *Commercial Vehicle* 12 (April 15, 1915): 21.
69. Joseph Husson, "Excessive Loading Time Decreases Efficiency of New York Brewery Trucks, Part IV," *Commercial Vehicle* 11 (December 15, 1914): 20–22. Although it is difficult to gauge the biases of the weighting system employed, it should be noted that electrics performed especially poorly on this combined score.
70. "Local Geography, Traffic, Trade and Paving Conditions Affect Delivery," *Commercial Vehicle* 13 (October 15, 1915): 18–24.
71. "Trucks Give Pittsburgh Department Stores Better Service at Less Cost," *Commercial Vehicle* 11 (August 15, 1914): 6.
72. Forty-five user reports collected from issues of *Commercial Vehicle* between January 1912 and June 1915.
73. The final report did conclude that "in the case of a large company whose operations include short, medium and long hauls, the greatest economy would probably be obtained by standardizing upon motor equipment rather than operating mixed service of horses and motors." See "Boston Tech. Fixes Fields of Motor Trucks," *Commercial Vehicle* 13 (November 1, 1915): 26–27.
74. La Schum, *The Electric Motor Truck,* 207, 218–21.
75. The is estimate based on a reported national fleet of 1225 electric trucks operating 310 days per year (ibid., 210).
76. "Gasoline and Electric Cost," *Commercial Vehicle* 7 (November 1912): 64.
77. "Selling the Electric Vehicle," *Commercial Vehicle* 7 (December 1912): 53.
78. Herbert Rice, discussing a paper by H. W. Hillman, in "Relative Importance of the Electric Truck As Compared with Other Classes of Central Station Business," *Central Station* 12 (February 1913): 258. Also see "Desirability of Battery Duplication," *Commercial Vehicle* 6 (April 1911): 236.
79. "Gas Cars in Central Station Use," *Electric Vehicles* 9 (October 1916): 137.
80. "You'll Not Catch Him Napping," *Commercial Vehicle* 13 (August 15, 1915): 13.
81. "Mileage in Wanamaker Service," *Commercial Vehicle* 6 (April 1911): 224.
82. "Efficient Truck Application Nets Cranford Co. 32.5 Per Cent. Saving," *Commercial Vehicle* 10 (May 1, 1914): 6–7.
83. "Makers Must Prepare Now," *Commercial Vehicle* 19 (August 15, 1918): 10.
84. "Use of Company Electric Vehicles by Electric Light Company," *Electrical World* 40 (November 1, 1902): 725.
85. "Central Station Automobiles," *Electrical World* 37 (February 23, 1901): 329.
86. "Hartford," *Horseless Age* 16 (July 5, 1905): 36.
87. "Recent Central Station Developments at Hartford, Conn.," *Electrical World* 52 (November 28, 1908): 1175–79; and "Commercial Central Station Practice at Hartford," *Electrical World* 56 (September 1, 1910): 489–91.
88. A. Jackson Marshall, "Battery Exchange System Applied to Electric Passenger Vehicles," *Central Station* 15 (October 1915): 103–5; Willis M. Thayer, "The Hartford Electric Light Company's Experience with the Battery Exchange System for Commercial Vehicles," *Central Station* 15 (December 1915): 141–44; and P. D. Wagoner, "Battery Service—A Unit in a Comprehensive Plan for the Successful Exploitation of the Electric Vehicle," *Proceedings of the National Electric Light Association, Thirty-ninth Annual Convention* (Chicago, June 22–26, 1916), 170–85.
89. P. D. Wagoner, commenting on "Report of the Committee on Electric Vehicles," in *Minutes of the Association of Edison Illuminating Companies* (annual meeting, 1914), 449.
90. "The Chicago Electric Vehicle Association," *Commercial Vehicle* 9 (November 1, 1913): 15; and Day Baker, "A Standardized Universal Battery and Compartment Essential for Making the Electric Truck Popular," *Central Station* 16 (February 1917): 179–80.
91. "Electric Vehicle Discussed," *Horseless Age* 2 (January 1897): 5; and Clinton Woods, "Future of the Self-propelled Vehicle," *Western Electrician* 21 (December 4, 1897): 315–17.

92. L. R. Wallis, "Central Stations and Automobiles," *Electrical World* 36 (December 15, 1900): 930–31.

93. W. H. Blood, "First Convention of the Electric Vehicle Association of America, President's Address," *Central Station* 10 (November 1910): 138.

94. "The Testimony of Users," *Commercial Vehicle* 2 (July 1907): 186; and William J. Johnson, "Battery Exchange Plan Prompts Motorization," *Commercial Vehicle* 13 (September 1, 1915): 9–11.

95. An article on electric vehicles in Philadelphia claimed that "a battery of, say, a dozen heavy trucks would warrant the establishment by the merchant . . . of a small repair plant, in the charge of a competent man, whose sole duty would be to keep the vehicles and their batteries constantly at their best." See "Experiences with Vehicles in Philadelphia," *Commercial Vehicle* 1 (April 1906): 68.

96. "Battery Service Progress Promises Electrics' Growth," *Commercial Vehicle* 15 (November 1, 1916): 45; and "Battery Service," *Electric Vehicles* 9 (October 1916): 147–49. A brief article reported growth in "installations of one, two, or three vehicles" (see "Electric Trucks in Hartford," *Commercial Vehicle* 10 [May 1, 1914]: 24). Also note that P. D. Wagoner reported that "smaller firms are adopting the electric ten times more rapidly where the system is available than where it is not" ("Battery Service," 181). Elsewhere Wagoner claimed that as much as 80 percent of electric vehicles sold on battery service were intended to replace gasoline trucks (see "Report of the Committee on Electric Vehicles," *Minutes of the Association of Edison Illuminating Companies* (annual meeting, 1914), 449.

97. "Rental Battery System," *Central Station* 15 (December 1915): 160.

98. "Battery Service Progress Promises Electrics' Growth," 45.

99. A Downing and Perkins mail truck was reported to have covered eighty-eight miles in a single day" (see Johnson, "Battery Exchange Plan," 11).

100. "Battery Exchange Service at Hartford," *Electrical World* 69 (April 28, 1917): 788–92.

101. Paul A. David has coined the term *angry orphans* to describe participants whose standards are not adopted. See his "Some New Standards for the Economics of Standardization in the Information Age," in *Economic Policy and Technological Performance,* ed. Partha Dasgupta and Paul Stoneman (New York: Cambridge University Press, 1987): 206–39.

102. "Hub Electric Men Protest Battery Plan," *Commercial Vehicle* 9 (September 1, 1915): 16–17.

103. "Vehicle Battery Service," *Commercial Vehicle* 9 (September 1, 1915): 12.

104. "Standardized Battery Service Recommended," *Electrical World* 69 (January 27, 1917): 166–67.

105. "Chicago Electric Truck Service Garage Finished," *Electrical World* 70 (December 29, 1917): 1247–48; "Vigorous Campaign on Electric Trucks in Chicago," *Electrical World* 71 (March 30, 1918): 676; and "New Electric Truck Service Ready in Chicago," *Electrical Review* 71 (December 29, 1917): 1091.

106. "Salvat Recommends New Milburn," *Electric Vehicles* 12 (January 1918): 22; "Chicago Service Plan for Electric Passenger Cars," *Electrical World* 69 (June 2, 1917): 1080; and "Making Batteries As Available As Gasoline," *Electrical World* 70 (July 28, 1917): 162. The original cost of the Milburn vehicle was reported as $1,685 in "A Complete List of American Electric Cars," *Electric Vehicles* 11 (November 1917): 137–38.

107. "Battery Exchange Service at Hartford," *Electrical World* 69 (April 28, 1917): 788–92.

108. "Developments in Electric Trucks at Hartford, Conn." *Electrical World* 73 (April 19, 1919): 796.

109. This was calculated from reported total mileage assuming the vehicles were used twenty-six days per month. At least five vehicles averaged forty or more miles per day, including the half-ton truck operated by the O.K. Bakery Company that covered forty-nine miles per day. Specific vehicles are listed in "Report of the Committee on Electric Vehicles" (1914), 441–42.

110. Note that the number of stalls reported includes all reported incidents, not simply those resulting from vehicles running out of power. See "Battery Exchange Service at Hartford," *Electrical World* 69 (April 28, 1917): 792.

111. This estimate assumes that the vehicle was in service six days per week (see ibid.).

112. "Battery Service Progress Promises Electrics' Growth," 45; and "Our Battery Department," *Illuminator* [Hartford] 1 (November 1922): 5.

113. "Hartford Electric Truck Service Plan," *Electrical World* 84 (October 18, 1924): 849–50.

114. "Electric Vehicles Contribute $400,000 to the Yearly Revenue of the Commonwealth Edison Company, Chicago," *Electrical World* 77 (May 14, 1921): 1096–97.

115. "New Battery Service Rate at Hartford," *Electrical World* 80 (October 7, 1922): 781.

116. "Results of French Trials," *Power Wagon* 2 (July 1908): 17.

117. "Conclusions from French Trials," *Commercial Vehicle* 5 (January 1910): 34–35; and "Russian Army Buying More Trucks," *Commercial Vehicle* 9 (January 1, 1914): 47.

118. "Opportunities for Demonstration," *Commercial Vehicle* 1 (July 1906): 176; and "No Trucks for Army Work," *Commercial Vehicle* 1 (August 1906): 213. In defense of the industry, the quartermaster issued the request on short notice, and most companies were already running their plants at full capacity.

119. "Motorized Warfare Will Modernize Transportation," *Commercial Vehicle* 11 (September 1, 1914): 14.

120. Raburn, "Motor Freight and Urban Morphogenesis," 168.

121. "Motor Preparedness," *Commercial Vehicle* 14 (February 1, 1916): 13.

122. "101 Makers Out of a Possible 200–Odd Bid on Five Classes of Army Trucks," *Commercial Vehicle* 16 (June 15, 1917): 14–15.

123. "Baker Electrics Invade England," *Commercial Vehicle* 11 (December 1, 1914): 11.

124. According to the 1920 census, 0.9 percent of commercial vehicles produced in 1919 (1,110 out of 126,436) were electric, while in 1909 electric vehicles had made up 10.9 percent (513 out of 4,700) of total production. See U.S. Bureau of the Census, *Fourteenth Census, 1920*, vol. 10: *Manufactures* (Washington, D.C.: Government Printing Office, 1921), 874.

125. Describing the situation in early 1916, another editorial noted: "We are not lacking in mileage of good roads. The difficulty is that such as we have are not correlated into any sort of logical or strategic system." See "Motor Preparedness," 13; and Raburn, "Motor Freight and Urban Morphogenesis," 183.

126. "Back to Business," *Commercial Vehicle* 19 (November 15, 1918): 26.

127. In "Motor Vehicle and Highway," 224, Romney reports that, by midsummer 1919, 27,983 trucks had been transferred to state ownership.

128. "Last Remnant of an Era," *Edison Round Table* (October 1947), 11; and memorandum from J. Mulholland, superintendent of transportation to J. F. Rice, June 6, 1950, Commonwealth Edison Company Library, Chicago, history folders, F3A/E3C; also see folder F32/E6.

5.   INFRASTRUCTURE, AUTOMOBILE TOURING, AND THE DYNAMICS OF AUTOMOTIVE SYSTEMS CHOICE

1. Albert Clough, "The Real Object of Their Hopes," *Horseless Age* 10 (August 20, 1902): 186.

2. "Holiday Festivities at the Boston Motor Club," *Electric Vehicles* 4 (January 1914): 17.

3. "Electric Vehicles in City Service," *Electrical World* 76 (August 28, 1920): 414–15.

4. Ibid.

5. "Record Run by Electric Roadster," *Electrical World* 62 (September 27, 1913): 617.

6. Fred B. Schaefer, "Colonel Bailey's 1,500 Mile Ride," *Electric Vehicles* 3 (November 1913): 245–47.

7. Ibid.

8. Maxwell G. Lay, *Ways of the World: A History of the World's Roads and of the Vehicles That Used Them* (New Brunswick, N.J.: Rutgers University Press, 1992), 5–6.

9. Bruce E. Seely, "Diffusion of Science into Engineering: Highway Research at the Bureau of Public Roads, 1900–1940," in *Transfer and Transformation of Ideas and Material Culture*, ed. Peter J. Hugill and D. Bruce Dickson (College Station: Texas A&M University Press, 1988), 144.

10. Norman T. Moline, *Mobility and the Small Town, 1900–1930: Transportation Change in Oregon, Illinois* (Chicago: University of Chicago, Department of Geography, 1971).

11. Peter J. Hugill, "Good Roads and the Automobile in the United States, 1880–1929," *Geographical Review* 72 (1982): 327–49.

12. Philip P. Mason, "The League of American Wheelmen and the Good Roads Movement, 1880–1905," (Ph.D. diss., University of Michigan, 1957), 185.

13. Wayne E. Fuller, *RFD: The Changing Face of Rural America* (Bloomington: Indiana University Press, 1964).

14. See, for example, National Highways Association, *National Highways Association Plan 1913*, Andrew Riker Papers, Bridgeport Public Library, Bridgeport, Conn.

15. Scholars with a range of historical interests have suggested that it was the advent of the automobile that finally stimulated demand for improved roads. See John C. Burnham, "The Gasoline Tax and the Automobile Revolution," *Mississippi Valley Historical Review* 98 (December 1961): 435–56; Paul Barrett, *The Automobile and Urban Transit: The Formation of Public Policy in Chicago, 1900–1930* (Philadelphia: Temple University Press, 1983); Moline, *Mobility and the Small Town;* and John B. Rae, *The American Automobile Industry* (Boston: Twayne, 1984).

16. "Road Improvement Fallacy," *Motor Age* 1, no. 6 (1899): 83.

17. R. H. Thurston, "Steam and Gas Engines on Automobiles," *Horseless Age* 7 (November 7, 1900): 40.

18. Arnulf Grübler, *The Rise and Fall of Infrastructures: Dynamics of Evolution and Technological Change in Transport* (Heidelberg, Germany: Physica-Verlag, 1990), 129, emphasis in original.

19. Charles Federic T. Young, *The Economy of Steam Power on Common Roads* (London: Atchley, 1860), 182.

20. The article also noted the importance of successful sales agents, the support of the local central station, and the prevalence of families owning more than one type of vehicle. See "How Much Is the 'Electric' Worth to a Community," *Electrical World* 70 (August 4, 1917): 212, emphasis added.

21. Historian James Laux attributes part of the persistence of the electric vehicle in the American marketplace to the state of American roads. See Jean-Pierre Bardou, Jean-Jacques Chanaron, Patrick Fridenson, and James M. Laux, *The Automobile Revolution: The Impact of an Industry,* trans. James M. Laux (Chapel Hill: University of North Carolina Press, 1982), 49.

22. Although the Stanley brothers hailed from Kingfield, Maine, the early success of steam in Portland cannot be attributed to the singular success of the Stanleys. Steam cars in Portland were evenly divided among the leading producers; 30 percent were Locomobiles (built from the basic Stanley design), 24 percent were Whites, 11 percent were Stanleys, and 13 percent were made either by the owner or other unknown producers. See "Minor Mention," *Horseless Age* 10 (December 3, 1902): 625.

23. This is based on an analysis of the *Vest Pocket Automobile Directory* (Cleveland: Jontzen, 1904; in the collection of the Western Reserve Historical Society, Cleveland). Data from 1904 show electric vehicles with 16 percent of the market and steam with 21 percent; numbers do not add up to 100 percent because mode of power was not reported in all instances. This analysis assumes that license numbers were issued chronologically.

24. "Licensed Automobile Operators in Chicago," *Western Electrician* 27 (September 15, 1900): 169.

25. "Peat As Fuel for Steam Vehicles," *Horseless Age* 7 (October 24, 1900): 16.

26. George Woodbury, *The Story of a Stanley Steamer* (New York: Norton, 1950), 64; and W. Brian Arthur, "Competing Technologies and Economic Prediction," *Options* (March 1984): 10–14. Abner Doble, builder of the Doble steam car, noted that a 95 percent efficient condenser allowed his vehicle to cover up to eight hundred miles on a single tank of water. See his "Steam Cars, Past and Present," *Automobile* 26 (November 2, 1916): 781.
    The outbreak of hoof-and-mouth disease has been misunderstood; Arthur has suggested the closing of the watering troughs was in part responsible for the failure of the steam car in competition with internal combustion. By the time condensers were fashioned in 1916 to allow the Stanley cars to tour without frequent water stops, the epidemic had so damaged the reputation of the steam vehicle that it never again seriously challenged the dominance of internal combustion. The unexpected outbreak of hoof-and-mouth disease inadvertently sealed the fate of the steam car, according to Arthur, and allowed the internal combustion engine to capture the entire automobile market. Although presented as part of an interesting and provocative theoretical inquiry into the nature of technological selection, Arthur's history of technology has the sequence of events wrong. By 1914 the technological competition to select the motive power for the passenger automobile had ended, and the internal combustion engine was already firmly established as the automotive standard. That year, while the Stanley brothers produced a grand total of 740 vehicles, Henry Ford produced a staggering 230,000 Model Ts. For the industry as a whole, the numbers were even more lopsided. The Census of Manufactures reports that 568,781 automobiles were produced in the United States. Only two firms reported making steam cars, and the Stanleys were the larger of the two; thus, production of steam vehicles was close to 1,000 units—less than 0.2 percent of the national total. See M. Mitchell Waldrop, *Complexity: The Science at the Edge of Order and Chaos* (New York: Simon & Schuster, 1992), 41; Robin Cowan, "Nuclear Power Reactors: A Study in Technological Lock-In," *Journal of Economic History* 50 (September 1990): 542–43; Charles C. McLaughlin, "The Stanley Steamer: A Study in Unsuccessful Innovation," in *Explorations in Enterprise,* ed. Hugh Aitken (Cambridge: Harvard University Press, 1965), 265–66; David A. Hounshell, *From the American System to Mass Production, 1800–1932: The Development of Manufacturing Technology in the United States* (Baltimore: Johns Hopkins University Press, 1984), 224; and U.S. Bureau of the Census, *Abstract of the Census of Manufactures, 1914* (Washington, D.C.: Government Printing Office, 1917), 225.

27. "Condensation in Steam Carriages," *Horseless Age* 9 (May 7, 1902): 547; E. J. Stoddard,

"Water for Automobile Boilers," *Horseless Age* 7 (December 26, 1900): 18–19; and "Recondensation of Steam in Steam Motor Vehicles," *Horseless Age* 5 (October 11, 1899): 9.

28. Hiram Percy Maxim, *Horseless Carriage Days* (New York: Dover, 1962), 37.

29. Albert Clough, "Bad Gasoline," *Horseless Age* 8 (April 24, 1901): 75–76.

30. James J. Flink, *America Adopts the Automobile, 1895–1910* (Cambridge: MIT Press, 1970), chap. 7. Also see Daniel Yergin, *The Prize: The Epic Quest for Oil, Money and Power* (New York: Simon & Schuster, 1991), part 1; and Stephen McIntyre, "'The Repair Man Will Gyp You': Mechanics, Managers and Customers in the Automobile Repair Industry" (Ph.D. diss., University of Missouri—Columbia, 1995), 60. On the availability of gasoline on Sundays, an article in *Horseless Age* asked rural motorists to encourage one local merchant or hotel operator in every town to sell gasoline on Sundays. See "Official Automobile Hotels," *Horseless Age* 7 (February 27, 1901): 20.

31. As David E. Nye observes, "utilities found that their technical and financial options pointed to street and commercial lighting in the 1880s, to electrical traction after 1888, to factories after the middle 1890s, to domestic business after 1910, and to farms only after 1935" (see his *Electrifying America* [Cambridge: MIT Press], 28).

32. Hiram Percy Maxim, "Radius of Action of Electric Motor Carriages," *Horseless Age* 2 (July 1897): 2–4.

33. Michael Brian Schiffer, *Taking Charge: The Electric Automobile in America* (Washington, D.C.: Smithsonian Institution Press, 1994), 65–66; and Hiram Percy Maxim, "Electric Vehicles and Their Relation to Central Stations," *Horseless Age* 3 (March 1899): 17–19.

34. Isaiah Roberts, "Facts about Storage Batteries," *Horseless Age* 4 (September 27, 1899): 9.

35. "Long Trip in an Electric Automobile," *Electrical Review* 53 (December 12, 1908): 909.

36. Harry E. Dey, "Electric Vehicle Charging—Charging Facilities in New York City," *Horseless Age* 16 (November 8, 1905): 608—9.

37. E. J. Bartlett, "Utility of Passenger and Commercial Electrics," *Electric Vehicles* 5 (July 1914): 3.

38. "The Central Station's Greatest Opportunity," *Electric Vehicles* 5 (September 1914): 99.

39. "Electrics in 1909 Glidden Tour," *Electrical Review* 54 (February 27, 1909): 402.

40. Electric vehicles did not participate in these events, although steam cars made by the White Motor Company excelled in these early reliability tests.

41. Justus B. Entz and Hiram Percy Maxim, "A Record Making Automobile," *Electrical World* 34 (December 23, 1899): 967–69. The two engineers claimed that the previous record had been 84.4 miles at an average speed of 11.6 miles per hour. At the conclusion of the run, the forty-eight cell battery registered ninety volts, indicating that it was nearly but not totally discharged.

42. "Storage Battery Automobiles" and "Electric Automobiling from Boston to New York," both in *Electrical World* 42 (October 31, 1903): 707, 726.

43. *Central Station* 9 (November 1909): 100.

44. "Claims Electric Mileage Record," *Horseless Age* 26 (November 14, 1910): 364.

45. In 1901 Frenchman Antoine Krieger claimed to have traveled 190.65 miles on a single charge, eclipsing the previous record of 187.5 set earlier that year in Cleveland. See "Electric Vehicle Runs 187 Miles on One Charge of Battery," *Western Electrician* 29 (August 17, 1901): 101; and "Automobiles," *Western Electrician* 29 (October 26, 1901): 283.

46. "Electric Automobiling from Boston to New York," *Electrical World* 42 (October 31, 1903): 726.

47. "Electrics Getting into the 'Touring' Class," *Central Station* 10 (September 1910): 78.

48. "The Methods of Automobile Manufacture," *Electrical World* 35 (May 26, 1900): 801.

49. "Some Dangers of Automobilism," *Electrical World* 38 (October 12, 1901): 580–81.

50. Quotations are from a transcript of Eames's statement and the introduction thereto. See "Mileage Capacity of Electric Vehicles," *Power Wagon* 2 (July 1908): 9.

51. "The Unimportance of Touring Ability," *Electric Vehicles* 9 (August 1916): 51–52.

52. Advertisement reprinted in *Electric Vehicles* 9 (August 1916): 50.

53. "The Unimportance of Touring Ability," 51–52.

54. Ibid.

55. A parallel for this pattern can be found in the current availability of two-wheel-drive sport utility vehicles.

EXCURSUS: THE ACCIDENTS THAT NEVER HAPPENED

1. Eames's remarks before the Association of Electric Vehicle Manufacturers are reprinted in "The Electric Wagon," *Power Wagon* 2 (April 1907): 11–12. Also see "Electric Vehicle for Business Purposes," *Commercial Vehicle* 2 (April 1907): 96–98; emphasis added.

2. Jon Elster, "Counterfactuals and the New Economic History," *Logic and Society* (New York: Wiley, 1978), 176–218; for "natural joint," see 180. For additional perspective on counterfactual theorizing, see Philip E. Tetlock and Aaron Belkin, eds., *Counterfactual Thought Experiments in World Politics* (Princeton: Princeton University Press, 1996); and Neal J. Reese and James M. Olson, eds., *What Might Have Been: The Social Psychology of Counterfactual Thinking* (Mahwah, N.J.: Erlbaum, 1995).

6.   THE BURDEN OF HISTORY

1. "The Electric Vehicle Situation," *Electrical World* 53 (March 4, 1909): 534.
2. One exception to this statement is the industrial electric truck, which did succeed in transforming freight and baggage handling within piers, rail stations, warehouses, and manufacturing facilities.
3. Nathan Rosenberg, "On Technological Expectations," in *Inside the Black Box: Technology and Economics* (New York: Oxford University Press, 1982), 104–19.
4. Harold Passer, *The Electrical Manufacturers, 1875–1900* (Cambridge: Harvard University Press, 1953), 45, quoted in Rosenberg, "On Technological Expectations," 112.
5. William J. Abernathy, *The Productivity Dilemma* (Baltimore: Johns Hopkins University Press, 1978); William J. Abernathy and James M. Utterback, "Patterns of Industrial Innovation," *Technology Review* 80 (June–July 1978): 40–47; Eric von Hippel, "Lead Users: A Source of Novel Product Concepts," *Management Science* 32 (July 1986): 791–805; and Geoffrey Moore, *Crossing the Chasm: Marketing and Selling Technology Products to Mainstream Customers* (New York: HarperBusiness, 1991).
6. The digital watch time line is adapted from Lemelson Center, "The Quartz Watch" (online exhibit), http://www.si.edu/lemelson/Quartz/.
7. Gijs Mom, "Inventing the Miracle Battery: Thomas Edison and the Electric Vehicle" (unpublished paper, 1998).
8. "Horseless Carriages," *Electrical World* 29 (April 10, 1897): 468.
9. "The Electric Motor Carriage," *Electrical World* 29 (May 15, 1897): 608.
10. "Electromobile Accumulators," *Electrical Review* 34 (June 21, 1899): 390.
11. "Electromobile," *Electrical Review* 37 (September 26, 1900): 301.
12. "Public Electric Vehicle Service," *Electrical Review* 38 (May 11, 1901): 571.
13. "Electrical Vehicle to the Front," *Electrical World* 56 (December 8, 1910): 1339–40.
14. "The Electric Vehicle," *Electrical World* 55 (June 16, 1910): 1580–81.
15. "Central Station Activity in the Electric Vehicle Field," *Electrical World* 58 (December 23, 1911): 1523–24.
16. "The Electric Vehicle," *Electrical World* 55 (January 6, 1910): 11–12.
17. "Thinking Straight on the Electric Vehicle," *Electrical World* 69 (February 24, 1917): 353.
18. "Electrics," in "Central-Station Service," *Electrical World* 84 (August 23, 1924): 365–67.
19. Mom, "Inventing the Miracle Battery," 8–9.
20. "Electric Vehicles," *Electrical World* 53 (March 25, 1909): 717.
21. Michael Brian Schiffer, *Taking Charge: The Electric Automobile in America* (Washington, D.C.: Smithsonian Institution Press, 1994), 153–55.
22. Ford electric vehicle 1914 file, Research Center, Henry Ford Museum and Greenfield Village, Dearborn, Mich.
23. Louis Bell, "Accumulators and Automobiles," *Electrical Review* 38 (April 27, 1901): 524; and "Electrical Vehicle to the Front," *Electrical World* 56 (December 8, 1910): 1339–40.
24. The editorial "In the Matter of Automobilism," *Electrical World* 36 (September 15, 1900): 400.
25. "Promoting Interest in Electric Vehicles," *Horseless Age* 26 (October 19, 1910): 531.
26. "Long Distance High Speed Electric Passenger Vehicle Is Here," *Central Station* 11 (March 1912): 268.
27. "Mr. Glidden As an Automobilist," *Western Electrician* 29 (July 13, 1901): 18.
28. W. H. Palmer, "The Storage Battery in the Commercial Operation of Electric Automobiles," *Electrical World* 39 (April 12, 1902): 646.
29. Discussion of 1912 data is reported in *Electric Vehicles* 4 (February 1914): 65.
30. U.S. Department of Commerce, Panel on Electrically Powered Vehicles, *The Automobile and Air Pollution* (Washington, D.C.: Government Printing Office, 1967), 2.
31. *Product Engineering* 38 (July 1967): 24. Walter Hayes, a spokesman for Ford of Britain, made this claim even more explicitly. Discussing their prototype Comuta vehicle, a two-person micro-car, Hayes stated: "Unless it can match the range and performance of a car like the Ford Galaxie," the electric car doesn't stand "much of a chance in the U.S." (*Product Engineering* 39 [April 1968]: 16).

32. Numerous companies announced the introduction of electric vehicles. By the standards of corporate America, none was ever mass-produced. The Sebring-Vanguard Citicar, however, was marketed nationally between 1973 and 1976, with sales totaling approximately 2,250 vehicles. See Robert G. Beaumont, "When America Made Electric Cars," *Earth Island Journal* 7 (Spring 1992): 39. Also see the table "Representative Electric Cars, 1967–77," in William Hamilton, *Electric Automobiles: Energy, Environmental, and Economic Prospects for the Future* (New York: McGraw-Hill, 1980), 8.

33. U.S. Senate, Committee on Commerce, *Electric Vehicle Research, Development, and Demonstration Act of 1975* (Washington, D.C.: General Printing Office, 1975), 40.

34. Ibid., 42.

35. Ibid., 136.

36. *Machine Design* 46 (October 17, 1974): 115.

37. *Electric Vehicle News* 7 (August 1978): 26.

38. Legislative language is quoted in *Electric Vehicle News* 7 (August 1978): 25. For a full report on the electric test vehicle, see U.S. Department of Energy, *Near Term Electric Test Vehicle: Phase II—Final Report*, DOE/CS/51294–01 (Washington, D.C.: Government Printing Office, October 1980), E1–E15. On test-tube batteries, see Hamilton, *Electric Automobiles*, 80–81. On development of propulsion, controller, and drive train subsystems, see U.S. Department of Energy, Energy Research and Development Administration, *Electric Vehicle Systems, FY1978*, EDP/C-01 (Washington, D.C.: Government Printing Office, August 1977), 22–23.

39. Gil Andrews Pratt, "EVs: On the Road Again," *Technology Review* 95 (August–September 1992): 52.

40. After some political wrangling, it was determined that only electric vehicles would count as ZEVs for the purposes of the mandate. There are, of course, emissions associated with electric vehicle use, but electrics are the only class of vehicle with zero tailpipe emissions.

41. Allen J. Scott, ed., *Electric Vehicle Manufacturing in Southern California: Current Developments, Future Prospects* (Los Angeles: University of California, Los Angeles; Lewis Center for Regional Policy Studies, June 1993).

42. David Frum, William Kristol, Frank Luntz, and Paul Tough, "A Revolution or Business As Usual?" *Harper's* 290 (March 1995): 43–53.

43. Lester B. Lave, Chris T. Hendrickson, and Francis Clay McMichael, "Environmental Implications of Electric Cars," *Science* 268 (May 19, 1995): 993–95; and Richard de Neufville, Stephen R. Connors, Frank R. Field III, David Marks, Donald R. Sadoway, and Richard D. Tabors, "The Electric Car Unplugged," *Technology Review* 99 (January 1996): 50–59.

44. On the history of EV1, see Michael Shnayerson, *The Car That Could: The Inside Story of GM's Revolutionary Electric Vehicle* (New York: Random House, 1996). Also see the General Motors press kit of January 4, 1996, especially "General Motors Electric Vehicle Historical Summary," in the collection of the author.

45. Schiffer, *Taking Charge*, 183.

46. This consortium, created in October 1991, committed the federal government and the U.S. automobile industry to spending $260 million over five years to support the development of an advanced battery suitable for use in electric passenger vehicles. See the Electric Power Research Institute, "Technical Brief RP3305–1" (July 1992).

47. F. R. Kalhammer, A. Kozawa, C. B. Moyer, and B. B. Owens, *Performance and Availability of Batteries for Electric Vehicles: A Report of the Battery Technical Advisory Panel* (El Monte: California Air Resources Board, December 1995).

7.   TECHNOLOGICAL HYBRIDS AND THE AUTOMOBILE SYSTEM

1. "The Electrified Gas Car," *Electric Vehicles* 10 (February 1917): 60.

2. Jesse H. Ausubel and Cesare Marchetti, "Elektron: Electrical Systems in Retrospect and Prospect," *Daedalus* 125 (Summer 1996): 155.

3. Calvin C. Burwell, "Transportation: Electricity's Changing Importance over Time," in *Electricity in the American Economy: Agent of Technological Progress*, ed. Sam H. Schurr, Calvin C. Burwell, Warren D. Devine, and Sidney Sonenblum (New York: Greenwood, 1990), 209–31.

4. Etymologically, the term *hybrid* has historically been associated with breeding: *Webster's* defines a hybrid as "an offspring of two animals or plants of different races, breeds, varieties, species, or genera." Because students of technological change continue to borrow methods and ideas from evolutionary theories of biological change, we can envision the general notion of a technological hybrid: an artifact or system that combines attributes of

two or more preceding technological generations. The conceptual leap to technological hybridization does not sidestep the many practical and epistemological problems inherent in the translation of evolutionary methods from the natural to the social world. Defining the boundaries and attributes of technology will always be fraught with challenges that are absent in the natural sciences. Fundamental distinctions such as phenotype and genotype pose serious problems in the domain of technology. Nonetheless, this chapter proceeds in the belief that the benefits and explanatory power of the idea of hybridization—defined as any technological system or artifact that combines discrete, preexisting systems, configurations, or artifacts to create new functional capabilities—outweigh these potential pitfalls. See Joel Mokyr, *The Lever of Riches: Technological Creativity and Economic Progress* (New York: Oxford University Press, 1990), chap. 11; George Basalla, *The Evolution of Technology* (New York: Cambridge University Press, 1988); and Robert Boyd and Peter Richerson, *Culture and the Evolutionary Process* (Chicago: University of Chicago Press, 1985).

5. Michael B. Schiffer first stimulated this line of thinking when he referred to the standard American automobile as an "electrified gasoline car." As I note later in this chapter, editorialists at *Electric Vehicles* in 1917 made a similar observation, referring instead to the "electrified gas car." See Schiffer, "Social Theory and History in Behavioral Archaeology," in *Expanding Archaeology*, ed. James M. Skibo, William H. Walker, and Axel E. Nielsen (Salt Lake City: University of Utah Press, 1995), 22–35; and "The Electrified Gas Car," *Electric Vehicles* 10 (February 1917): 60.

6. The story of the starter motor is not complete; Kettering's was only one possible solution to the starter problem.

7. Richard Schallenberg, *Bottled Energy* (Philadelphia: American Philosophical Society, 1982), 240; and John P. McKay, *Tramways and Trolleys: The Rise of Urban Mass Transport in Europe* (Princeton: Princeton University Press, 1976), 74–78.

8. In Germany the hybrid wagons were called "Güterwagen für Land und Schweinwege." See Woolsey M. Johnson, "The Street Railway System of Hanover, Germany," *Electrical World* 34 (October 14, 1899): 572–73.

9. E. Hospitalier, "Self-Moving Road Vehicles with Automobile Double Trolleys," *Electrical Review* 36 (March 7, 1900): 230–31; and "Trolley-Fed Automobiles in France," *Electrical World* 35 (March 10, 1900): 358–59. Lacking a conductive rail, the double-trolley design was necessary to create a complete electrical circuit; most traditional trolley design used the railway as the return.

10. R. W. Shoemaker, "The Trackless Trolley at Los Angeles," *Electrical World* 56 (October 27, 1910): 1002–3.

11. The "Laurel Canyon" entry in Leonard Pitt and Dale Pitt, eds., *Los Angeles A to Z: Encyclopedia of the City and County* (Los Angeles: University of California Press, 1997).

12. "American Electro-Gasoline Buses for London Streets," *Horseless Age* 10 (July 9, 1902): 76–77; "R. H. Macy's & Co's Combination Truck," *Horseless Age* 10 (September 24, 1902): 345; and "Combination Electric Automobile," *Electrical World* 39 (February 15, 1902): 317.

13. "List of Entries for the 100–mile Endurance Run of the Automobile Club of America," *Horseless Age* 9 (May 28, 1902): 663. Note that the Vehicle Equipment Company was an orphan of the EVC conglomerate and later became the General Vehicle Company, one of the largest manufacturers of electric commercial vehicles in the 1910s.

14. The Narragansett races ran on September 24, 1902. See *Horseless Age* 10 (October 1, 1902): 355; and "General Table of Entries," *Horseless Age* 10 (October 15, 1902): 410–11.

15. "Two Unique Gasolene Electric Auto-Trucks," *Electrical Review* 54 (February 13, 1909): 300–301; and "The Couple Gear Gas-Electric Truck," *Horseless Age* 24 (August 25, 1909): 214.

16. "Gasoline-Electric Express Wagon for Suburban Service," *Electrical World* 54 (December 30, 1909): 1586; "Gasoline-Electric Truck," *Electrical World* 54 (August 19, 1909): 444; and "Gasoline-Electric Passenger Vehicle, *Electrical World* 56 (August 11, 1910): 346–47.

17. "Two-Passenger Gasoline-Electric Automobile," *Electrical World* 65 (April 3, 1915): 870; and "Gasoline Electric Automobile," *Electrical World* 65 (February 6, 1915): 358.

18. "Economical Electric-Gasoline Car," *Electrical World* 67 (March 25, 1916): 730.

19. "Electric Vehicle Manufacturer to Build Gasoline-Electric Car with Magnetic Clutch," *Electrical World* 66 (December 25, 1915): 1447; and "Electric Transmission Increases Flexibility of Gasoline Operated Automobiles," *Electrical World* 67 (January 8, 1916): 110. For more on the Owen Magnetic, see Richard Wager, *Golden Wheels: The Story of the Automobiles Made in Cleveland and Northeastern Ohio, 1892–1932* (Cleveland: Western Reserve Historical Society, 1986).

20. According to Schallenberg, the electric starter was the "nail in the coffin" of the electric car (see *Bottled Energy*, 274). Also see Maxwell G. Lay, *Ways of the World: A History of the*

World's Roads and of the Vehicles That Used Them (New Brunswick, N.J.: Rutgers University Press, 1992): 163; Virginia Scharff, "Gender, Electricity, and Automobility," in *The Car and the City: The Automobile, The Built Environment, and Daily Urban Life,* ed. Martin Wachs and Margaret Crawford (Ann Arbor: University of Michigan Press), 82–83; and T. A. Boyd, "The Self-Starter," *Technology and Culture* 9 (July 1968): 585–91.

21. "The Starting of Gasoline Motors," *Horseless Age* 9 (March 19, 1902): 353.
22. "A.C.A. Commercial Vehicle Contest," *Horseless Age* 11 (May 27, 1903): 626.
23. This formulation was suggested by Thomas Turrentine and Kenneth Kurani, "The Household Market for Electric Vehicles" (working paper, Institute of Transportation Studies, University of California, Davis, May 12, 1995).
24. *Vest Pocket Automobile Directory* (Cleveland: Jontzen, Cleveland, 1904, in the collection of the Western Reserve Historical Society, Cleveland).
25. "The Electric Automobile in Paris," *Electrical World* 37 (January 12, 1901): 102.
26. In 1895 *Horseless Age* claimed that Dr. Carlos C. Booth of Youngstown, Ohio, was the first physician-motorist, and testimonials to the merits of different types of vehicles appear regularly after 1901. See the section "Doctor's Automobiles," *Horseless Age* 7 (February 6, 1901): 21–38. Also see *Horseless Age* 8 (November 6, 1901): 657–71; and 11 (January 7, 1903): 3–14.
27. Virginia Scharff, *Taking the Wheel: Women and the Coming of the Motor Age* (New York: Free Press, 1991); and Gijs Mom, "Electric Vehicles: Male or Female?" (paper presented at the Intercontinental Mobile Sources Clean Air Conference, Munich, March 1996).
28. "The 'Cosmopolitan' Race," *Horseless Age,* 1 (June 1896): 5–6.
29. "Minor Mention," *Horseless Age* 6 (June 6, 1900): 25; "Excluded from Chicago Parks," *Horseless Age* 4 (June 21, 1899): 6; "Park Exclusion Rule Contested," *Horseless Age* 5 (November 1, 1899): 6–7; "Barred from Golden Gate Park," *Horseless Age* 5 (March 7, 1900): 19; and "Park Exclusion Case to Be Tested at Baltimore," *Horseless Age* 7 (January 9, 1901): 13.
30. "More Park Permits" *Horseless Age* 5 (December 13, 1899): 8; and "Minor Mention," *Horseless Age* 5 (March 21, 1900): 17.
31. W. M. Hutchison, M.D.,"The Objection to Pasted Plates," *Horseless Age* 6 (August 1, 1900): 15.
32. "Park Exclusion Case to Be Tested at Baltimore," 13.
33. "Minor Mention," *Horseless Age* 9 (April 15, 1902): 460.
34. "Vehicle Garaging in London," *Electric Vehicles* 5 (August 1914): 46.
35. "Electrics by Compulsion," *Electric Vehicles* 10 (April 1917): 131.
36. Lay, *Ways of the World,* 199.
37. Ross D. Eckert and George W. Hilton, "The Jitneys," *Journal of Law and Economics* 15, no. 2 (1972): 293–325.
38. "Electric Machinery on Automobiles," *Electric Vehicles* 3 (December 1913): 298.
39. "Automotive Association Holds First Meeting," *Electric Vehicles* 10 (June 1917): 193.
40. "National Automotive Electric Service Association," *Electrical World* 72 (December 21, 1918): 1195; "Electrical Automobile Accessories," *Electric Vehicles* 11 (September 1917): 89–93; "Electric Automobile Heater," *Electrical World* 70 (October 20, 1917): 793; and "Electric Steering Wheel Warmer," *Electrical World* 70 (December 15, 1917): 1178.
41. "The Electrified Gasoline Automobile," *Electrical World* 61 (April 26, 1913): 861; and "The Automobile Slump," *Electrical World* 76 (October 23, 1920): 813.
42. Fred Allison, "The World's Largest Direct Current Station," *Electrical World* 68 (August 12, 1916): 312–16; and "Lighting in Automobile Factories," *Electrical World* 69 (June 30, 1917): 1253–55.
43. "Industrial Tractor Economies," *Electrical World* 76 (September 4, 1920): 481; and "Labor Economy of Industrial Trucks and Tractors," *Electrical World* 69 (June 30, 1917): 1250–52.
44. "Automobile Industry Highly Electrified," *Electrical World* 82 (September 22, 1923): 582.
45. Glenn Zorpette, "Alan Cocconi: Electric Cars and Pterosaurs Are My Business," *Scientific American* 276 (May 1997): 32–33.
46. Victor Wouk, "Hybrids: Then and Now," and Bradford Bates, "Getting a Ford HEV on the Road," both in *IEEE Spectrum* 32 (July 1995): 16–21, 21–25.
47. Electric Power Research Institute, *Building the Electric Vehicle Future: EPRI's Vehicle Development Activities* (Palo Alto, Calif.: Electric Power Research Institute, 1987).
48. This incremental approach runs counter to the hopes that electric vehicle enthusiasts of the 1960s and 1970s had for the wholesale replacement of internal combustion. The hybrid artifact has not been widely admired in engineering circles. The automotive mainstream saw hybrids as needlessly complicated; hybrid designs produced only marginal

improvements in efficiency at considerable cost. And for hobbyists and other industry outsiders, the hybrid was an "impure" technology that did not hold the same promise of transformation that the electric vehicle did. Electric vehicle supporters have been willing to tow an auxiliary generator to extend the range of their vehicles but unwilling to put both electric and internal combustion drive systems under one hood. For more on the culture of these hobbyists, see David A. Kirsch, "The Electric Vehicle and the Burden of History: Studies in Automotive Systems Rivalry in the United States, 1890–1996," (Ph.D. diss., Stanford University, 1996), chap. 6.

49. Turrentine and Kurani, "The Household Market for Electric Vehicles." According to the report, a small but significant group of drivers placed a high value on not having to go to the gas station to refuel their vehicle.

50. Barbara E. Taylor, *The Lost Cord: A Storyteller's History of the Electric Car* (Columbus, Ohio: Greydon, 1995).

51. See, for instance, proceedings of the Organization for Economic Cooperation and Development conference, *The Urban Electric Vehicle: Policy Options, Technology Trends, and Market Prospects* (Washington, D.C.: OECD, 1992).

52. On the general phenomenon of creating niche markets, see René Kemp, Johan Schot, and Remco Hoogma, "Regime Shifts to Sustainability through Processes of Niche Formation: The Approach of Strategic Niche Management," *Technology Analysis and Strategic Management* 10, no. 2 (1998): 175–95.

8.   INDUSTRIAL ECOLOGY AND THE FUTURE OF THE AUTOMOBILE

1. Albert L. Clough, "Automobiles and Sanitation," *Horseless Age* 24 (October 6, 1909): 368.

2. Nathan Rosenberg, *Perspectives on Technology* (New York: Cambridge University Press, 1976), 108–25; and Richard Nelson and Sidney Winter, *An Evolutionary Theory of Economic Change* (Cambridge: Belknap Press of Harvard University Press, 1982).

3. Paul A. David, "Some New Standards for the Economics of Standardization in the Information Age," in *Economic Policy and Technological Performance*, ed. Partha Dasgupta and Paul Stoneman (New York: Cambridge University Press, 1987), 206–39; and C. O. Quandt, "Manufacturing the Electric Vehicle: A Window of Technological Opportunity for Southern California," *Environment and Planning A* 27, no. 6 (1995): 835–62.

4. David, "Some New Standards," 214.

5. Clay McShane, *Down the Asphalt Path: The Automobile and the American City* (New York: Columbia University Press, 1994), 54.

6. R. Hendricks, *International Encyclopedia of Horse Breeds* (Norman: University of Oklahoma Press, 1995).

7. Peter Freund and George Martin, *The Ecology of the Automobile* (Cheektowaga, N.Y.: Black Rose Books, 1993), 9.

8. Perrin Meyer, "Bi-logistic Growth," *Technological Forecasting and Social Change* 47 (September 1994): 89–102.

9. On Haagen-Smit's pioneering work on the photochemical nature of smog, see James E. Krier and Edmund Ursin, *Pollution and Policy: A Case Essay on California and Federal Experience with Motor Vehicle Air Pollution, 1940–1975* (Los Angeles: University of California Press, 1977).

10. Alan P. Loeb, "Birth of the Kettering Doctrine: Fordism, Sloanism and the Discovery of Tetraethyl Lead," *Business and Economic History* 24, no. 1 (1995): 72–87; Jerome O. Nriagu, "The Rise and Fall of Leaded Gasoline," *Science of the Total Environment* 92 (March 1990): 13–28; and Nriagu, "A History of Global Metal Pollution," *Science* 272 (April 12, 1996): 223–24.

11. Jesse H. Ausubel and Hedy E. Sladovich, eds., *Technology and Environment* (Washington, D.C.: National Academy Press, 1989); and Deanna J. Richards, Braden R. Allenby, and Robert A. Frosch, "The Greening of Industrial Ecosystems: Overview and Perspective," in *The Greening of Industrial Ecosystems*, ed Braden R. Allenby and Deanna J. Richards (Washington, D.C.: National Academy Press, 1994), 3.

12. Thomas E. Graedel and Braden R. Allenby, *Industrial Ecology* (Englewood Cliffs, N.J.: Prentice Hall, 1995), 9.

13. Freund and Martin, *The Ecology of the Automobile;* also see Ann Y. Watson, Richard R. Bates, and Donald Kennedy, eds., *Air Pollution, the Automobile, and Public Health* (Washington, D.C.: National Academy Press, 1988).

14. Freund and Martin, *The Ecology of the Automobile,* 65–71.

15. René Kemp, "Environmental Policy and Technical Change: A Comparison of the Techno-

logical Impact of Policy Instruments" (Ph.D. diss., University of Limburg, Maastricht, the Netherlands, 1995); Johan W. Schot, "Constructive Technology Assessment and Technology Dynamics: The Case of Clean Technologies," *Science, Technology and Human Values* 17, no. 1 (1992): 36–56; and René Kemp, Johan Schot, and Remco Hoogma, "Regime Shifts to Sustainability through Processes of Niche Formation: The Approach of Strategic Niche Management," *Technology Analysis and Strategic Management* 10, no. 2 (1998): 175–95.

16. Stuart P. Beaton, Gary A. Bishop, Yi Zhang, Lowell L. Ashbaugh, Douglas R. Lawson, and Donald H. Stedman, "On-Road Vehicle Emissions: Regulations, Costs and Benefits," *Science* 268 (May 19, 1995): 991–93.

17. Lee Schipper, "Electric Vehicles in a Broader Context: Too Early or Too Late?" in *The Urban Electric Vehicle: Policy Options, Technology Trends, and Market Prospects,* ed. Organization for Economic Cooperation and Development (Stockholm: Organization for Economic Cooperation and Development, 1992), 381–85.

EPILOGUE

1. "December Meeting of the Electric Vehicle Association of America," *Central Station* 13 (January 1914): 291.

2. Edward Tenner, *Why Things Bite Back: Technology and the Revenge of Unintended Consequences* (New York: Knopf, 1996).

Bibliography

PRIMARY SOURCES

Archives

Boston Edison Company, Boston
    Periodicals and materials
Bridgeport Public Library, Bridgeport, Conn.
    Andrew S. Riker Papers
Columbia University, Rare Book and Manuscript Library, New York
    Richard W. Meade Papers
Commonwealth Edison Company Library, Chicago
    Periodicals and papers
Connecticut State Library and Archives, Hartford
    Hiram Percy Maxim Papers
Consolidated Edison Company Library, New York
    Periodicals and materials
Edison National Historic Site, West Orange, N.J.
    Thomas A. Edison Papers
Hartford Electric Light Company, Hartford
    Periodicals and materials
Henry Ford Museum and Greenfield Village Research Center, Dearborn, Mich.
    Frank C. Armstrong Papers
Library of Congress, Washington, D.C.
    William C. Whitney Papers
National Automotive History Collection, Detroit Public Library
    Henry Cave Papers
New Jersey Historical Society, Newark
    William F. D. Crane Papers
Studebaker National Historic Site, South Bend, Ind.
    Studebaker corporate archives
Western Reserve Historical Society, Cleveland
    White Motor Company bulletins
    Automotive history collection

Periodicals

American Machinist
Association of Edison Illuminating Companies, minutes
Automobile
Central Station
Commercial and Financial Chronicle
Commercial Vehicle
Electric Vehicle Association of America, proceedings
Electric Vehicle News
Electric Vehicles
Electrical Engineer
Electrical Review
Electrical World
Horseless Age
Illuminator
Journal of the Franklin Institute
Machine Design
Motor Age
Motor Traffic
National Electric Light Association Bulletin
National Electric Light Association, proceedings
New York Times
Power Wagon
Product Engineering
Western Electrician

SECONDARY SOURCES

Abernathy, William J. 1978. *The Productivity Dilemma: Roadblock to Innovation in the Automobile Industry.* Baltimore: Johns Hopkins University Press.

Abernathy, William J., and James M. Utterback. 1978. "Patterns of Industrial Innovation." *Technology Review* 80 (June–July): 40–47.

Adler, Sy. 1991. "The Transformation of the Pacific Electric Railway: Bradford Snell, Roger Rabbit, and the Politics of Transportation in Los Angeles." *Urban Affairs Quarterly* 27, no. 1: 51–86.

Anderson, Russell H. 1954. "First Automobile Race in America." *Illinois State Historical Society Journal* 47, no. 4: 343–59.

Arthur, W. Brian. 1984. "Competing Technologies and Economic Prediction." *Options* (March): 10–14.

———. 1988. "Competing Technologies: An Overview." In *Technical Change and Economic Theory,* edited by Giovanni Dosi, Christopher Freeman, Richard Nelson, Gerald Silverberg, and Luc Soete, 590–607. New York: Pinter.

Ausubel, Jesse H., and Cesare Merchetti. 1996. "Elecktron: Electrical Systems in Retrospect and Prospect." *Daedalus* 125, no. 3: 139–69.

Ausubel, Jesse H., and Hedy E. Sladovich, eds. 1989. *Technology and Environment.* Washington, D.C.: National Academy Press.

Bacon, John H. 1984. *American Steam-Car Pioneers: A Scrapbook.* Exton, Pa.: Newcomen Society of the United States.

Baldwin, Neil. 1995. *Edison: Inventing the Century.* New York: Hyperion.

Bardou, Jean-Pierre, Jean-Jacques Chanaron, Patrick Fridenson, and James M. Laux. 1982. *The Automobile Revolution: The Impact of an Industry,* translated by James M. Laux. Chapel Hill: University of North Carolina Press.

Barker, Theo, ed. 1987. *The Economic and Social Effects of the Spread of Motor Vehicles: An International Centenary Tribute.* London: Macmillan.

———. 1993. "Slow Progress: Forty Years of Motoring Research." *Journal of Transport History,* 3d series, 14 (September): 142–65.

Barrett, Paul. 1983. *The Automobile and Urban Transit: The Formation of Public Policy in Chicago, 1900–1930.* Philadelphia: Temple University Press.

Basalla, George. 1988. *The Evolution of Technology.* New York: Cambridge University Press.

Bates, Bradford. 1995. "Getting an SUV on the Road." *IEEE Spectrum* 32 (July): 21–25.

Beasely, David. 1988. *The Suppression of the Automobile: Skulduggery at the Crossroads*. New York: Greenwood.

Beaton, Stuart P., Gary A. Bishop, Yi Zhang, Lowell L. Ashbaugh, Douglas R. Lawson, and Donald H. Stedman. 1995. "On-Road Vehicle Emissions: Regulations, Costs and Benefits." *Science* 268 (May 19): 991–93.

Beaumont, Robert G. 1992. "When Americans Made Electric Cars." *Earth Island Journal* 7 (Spring): 39.

Belasco, Warren James. 1997 [1979]. *Americans on the Road: From Autocamp to Motel, 1910–1945*. Baltimore: John Hopkins University Press.

Berger, Michael L. 1979. *The Devil Wagon in God's Country: The Automobile and Social Change in Rural America, 1893–1929*. Hamden, Conn.: Archon.

Berger, Peter L., and Thomas Luckmann. 1966. *The Social Construction of Reality: A Treatise in the Sociology of Knowledge*. Garden City, N.Y.: Doubleday.

Bottles, Scott L. 1987. *Los Angeles and the Automobile: The Making of the Modern City*. Berkeley: University of California Press.

Boyd, Robert, and Peter J. Richerson. 1985. *Culture and the Evolutionary Process*. Chicago: University of Chicago Press.

Boyd, T. A. 1968. "The Self-Starter." *Technology and Culture* 9 (July): 585–91.

Bradley, James, and Richard Langworth. 1971. "Calendar Year Production: 1896 to Date." In *The American Car Since 1775*, edited by *Automobile Quarterly*, 138–43. New York: Dutton.

Burnham, John C. 1961. "The Gasoline Tax and the Automobile Revolution." *Mississippi Valley Historical Review* 98 (December): 435–56.

Burwell, Calvin C. 1990. "Transportation: Electricity's Changing Importance over Time." In *Electricity in the American Economy: Agent of Technological Progress*, edited by Sam H. Schurr, Calvin C. Burwell, Warren D. Devine, and Sidney Sonenblum. New York: Greenwood.

Callon, Michel. 1980. "The State and Technical Innovation: A Case Study of the Electrical Vehicle in France." *Research Policy* 9 (May–June): 358–76.

———. 1987. "Society in the Making: The Study of Technology As a Tool for Sociological Analysis." In *The Social Construction of Technological Systems*, edited by Wiebe E. Bijker, Thomas P. Hughes, and Trevor J. Pinch, 83–103. Cambridge: MIT Press.

Carman, Harry James. 1969 [1919]. *The Street Surface Railway Franchises of New York City*. New York: AMS.

Cheape, Charles W. 1980. *Moving the Masses: Urban Public Transit in New York, Boston, and Philadelphia, 1880–1912*. Cambridge: Harvard University Press.

Cowan, Robin. 1990. "Nuclear Power Reactors: A Study In Technological Lock-In." *Journal of Economic History* 50 (September): 541–67.

Cudahy, Brian J. 1990. *Cash, Tokens, and Transfers: A History of Urban Mass Transit in North America*. New York: Fordham University Press.

David, Paul A. 1985. 'Clio and the Economics of QWERTY.' *American Economic Review* 75 (May): 332–37.

———. 1987. 'Some New Standards for the Economics of Standardization in the Information Age.' In *Economic Policy and Technological Performance*, edited by Partha Dasgupta and Paul Stoneman, 206–39. New York: Cambridge University Press.

———. 1992. "Heroes, Herds and Hysteresis in Technological History: Thomas Edison and 'The Battle of the Systems' Reconsidered." *Industrial and Corporate Change* 1, no. 1: 129–80.

———. 1997. *Path Dependence and the Quest for Historical Economics: One More Chorus of the Ballad of QWERTY*. Discussion Papers in Economic and Social History 20. Oxford: University of Oxford.

de Neufville, Richard, Stephen R. Connors, Frank R. Field III, David Marks, Donald R. Sadoway, and Richard D. Tabors. 1996. "The Electric Car Unplugged." *Technology Review* 99, no. 1: 30–36.

Doolittle, James Rood. 1916. *The Romance of the Automobile Industry*. New York: Klebold.

Duffy, M. C. 1995. "The Mercury-Arc Rectifier and Supply to Electric Railways." *Engineering Science and Education Journal* 4 (August): 183–92.

Eckert, Ross D., and George W. Hilton. 1972. "The Jitneys." *Journal of Law and Economics* 15, no. 2: 293–325.

Electric Power Research Institute. 1987. *Building the Electric Vehicle Future: EPRI's Vehicle Development Activities*. Palo Alto, Calif.: Electric Power Research Institute.

———. 1992. "Technical Brief RP3305–1." July.

Elster, Jon. 1978. *Logic and Society*. New York: Wiley.

Fischer, Claude S. 1992. *America Calling: A Social History of the Telephone to 1940*. Berkeley: University of California Press.

Flink, James J. 1970. *America Adopts the Automobile, 1895–1910*. Cambridge: MIT Press.

———. 1980. "The Car Culture Revisited: Some Comments on the Recent Historiography of Automotive History." *Michigan Quarterly Review* 19 (Fall–Winter): 772–81.

———. 1988. *The Automobile Age*. Cambridge: MIT Press.

Foster, Mark. 1981. *From Streetcar to Superhighway: American City Planners and Urban Transportation, 1900–1940*. Philadelphia: Temple University Press.

Freund, Peter, and George Martin. 1993. *The Ecology of the Automobile*. Cheektowaga, N.Y.: Black Rose Books.

Frum, David, William Kristol, Frank Luntz, and Paul Tough. 1995. "A Revolution or Business As Usual?" *Harper's* 290 (March): 43–53.

Fuller, Wayne. 1964. *RFD: The Changing Face of Rural America*. Bloomington: University of Indiana Press.

Funnell, Charles E. 1975. *By the Beautiful Sea: The Rise and High Times of That Great American Resort, Atlantic City*. New York: Knopf.

Gilbert, Gorman, and Robert E. Samuels. 1982. *The Taxicab: An Urban Transportation Survivor*. Chapel Hill: University of North Carolina Press.

Goddard, Stephen B. 1944. *Getting There: The Epic Struggle between Road and Rail in the American Century*. Chicago: University of Chicago Press.

Graedel, Thomas E., and Braden R. Allenby. 1995. *Industrial Ecology*. Englewood Cliffs, N.J.: Prentice Hall.

Grübler, Arnulf. 1990. *The Rise and Fall of Infrastructures: Dynamics of Evolution and Technological Change in Transport*. Heidelberg, Germany: Physica-Verlag.

Hamilton, William. 1980. *Electric Automobiles: Energy, Environmental, and Economic Prospects for the Future*. New York: McGraw-Hill.

Hammack, David C. 1987. *Power and Society: Greater New York at the Turn of the Century*. New York: Columbia University Press.

Hausman, William J., and John L. Neufeld. 1990. "The Structure and Profitability of the US Electric Utility Industry at the Turn of the Century." *Business History* 32 (April): 225–43.

Hendricks, R. 1995. *International Encyclopedia of Horse Breeds*. Norman: University of Oklahoma Press.

Hirsch, Mark D. 1948. *William C. Whitney: Modern Warwick*. New York: Dodd, Mead.

Hirsh, Richard F. 1989. *Technology and Transformation in the American Utility Industry*. New York: Cambridge University Press.

Hounshell, David A. 1984. *From the American System to Mass Production, 1800–1932: The Development of Manufacturing Technology in the United States*. Baltimore: Johns Hopkins University Press.

Hughes, Thomas P. 1971. *Elmer Sperry: Inventor and Engineer*. Baltimore: Johns Hopkins University Press.

———. 1983. *Networks of Power: Electrification in Western Society, 1880–1930*. Baltimore: Johns Hopkins University Press.

———. 1987. "The Evolution of Large Technological Systems." In *The Social Construction of Technological Systems*, edited by Wiebe E. Bijker, Thomas P. Hughes, and Trevor Pinch, 51–82. Cambridge: MIT Press.

———. 1989. *American Genesis: A Century of Invention and Technological Enthusiasm*. New York: Penguin.

———. 1994. "Technological Momentum." In *Does Technology Drive History? The Dilemma of Technological Determinism*, edited by Merritt Roe Smith and Leo Marx, 101–14. Cambridge: MIT Press.

Hugill, Peter J. 1982. "Good Roads and the Automobile in the United States, 1880–1929." *Geographical Review* 72: 327–49.

———. 1988. "Technology Diffusion in the World Automobile Industry, 1885–1985." In *The Transfer and Transformation of Ideas and Material Culture*, ed. Peter J. Hugill and D. Bruce Dickson, 110–62. College Station: Texas A&M University Press.

Irvine, Andrew C. 1954. "The Promotion and First Twenty-Two Years' History of a Corporation in the Electrical Manufacturing Industry." M.A. dissertation, Temple University.

Jackson, Kenneth T. 1985. *Crabgrass Frontier: The Suburbanization of the United States*. New York: Oxford University Press.

Jakle, John A., and Keith A. Sculle. 1994. *The Gas Station in America*. Baltimore: Johns Hopkins University Press.

Jamison, Andrew. 1970. *The Steam-Powered Automobile: An Answer to Air Pollution.* Bloomington: Indiana University Press.

Joerges, Bernward. 1988. "Large Technical Systems: Concepts and Issues." In *The Development of Large Technical Systems,* edited by Renate Mayntz and Thomas P. Hughes, 9–36. Boulder, Colo.: Westview.

Kalhammer, F. R., A. Kozawa, C. B. Moyer, and B. B. Owens. 1995. *Performance and Availability of Batteries for Electric Vehicles: A Report of the Battery Technical Advisory Panel.* El Monte, Calif.: California Air Resources Board.

Kemp, René. 1995. "Environmental Policy and Technical Change: A Comparison of the Technological Impact of Policy Instruments." Ph.D. dissertation, University of Limburg, Maastricht, the Netherlands.

Kemp, René, Johan Schot, and Remco Hoogma. 1998. "Regime Shifts to Sustainability through Processes of Niche Formation: The Approach of Strategic Niche Management." *Technology Analysis and Strategic Management* 10, no. 2: 175–95.

Kirsch, David A. 1996. "The Electric Vehicle and the Burden of History: Studies in Automotive Systems Rivalry in the United States, 1980–1996." Ph.D. dissertation, Stanford University.

Kline, Ronald R. 1992. *Steinmetz: Engineer and Socialist.* Baltimore: Johns Hopkins University Press.

Konvitz, Josef, Mark Rose, and Joel Tarr. 1990. "Technology and the City." *Technology and Culture* 20 (April): 284–94.

Krier, James E., and Edmund Ursin. 1977. *Pollution and Policy: A Case Essay on California and Federal Experience with Motor Vehicle Pollution, 1940–1975.* Los Angeles: University of California Press.

La Schum, Edward E. 1924. *The Electric Motor Truck: Selection of Motor Vehicle Equipment, Its Operation and Maintenance.* New York: U.P.C. Book Company.

Latour, Bruno. 1987. *Science in Action: How to Follow Scientists and Engineers through Society.* Cambridge: Harvard University Press.

———. 1992. "Where Are the Missing Masses? The Sociology of a Few Mundane Artifacts." In *Shaping Technology/Building Society: Studies in Sociotechnical Change,* edited by Wiebe E. Bijker and John Law, 225–58. Cambridge: MIT Press.

Lave, Lester B., Chris T. Hendrickson, and Francis Clay McMichael. 1995. "Environmental Implications of Electric Cars." *Science* 268 (May 19): 993.

Law, John. 1987. "Technology and Heterogeneous Engineering: The Case of Portuguese Expansion." In *The Social Construction of Technological Systems,* edited by Wiebe E. Bijker, Thomas P. Hughes, and Trevor Pinch, 111–34. Cambridge: MIT Press.

Lay, Maxwell G. 1992. *Ways of the World: A History of the World's Roads and of the Vehicles That Used Them.* New Brunswick, N.J.: Rutgers University Press.

Lemonnier, Pierre, ed. 1993. *Technological Choices: Transformation in Material Culture Since the Neolithic.* New York: Routledge.

Levine, Gary. 1974. *The Car Solution: The Steam Engine Comes of Age.* New York: Horizon.

Loeb, Alan P. 1995. "Birth of the Kettering Doctrine: Fordism, Sloanism and the Discovery of Tetraethyl." *Business and Economic History* 24, no. 1: 72–87.

McIntyre, Stephen. 1995. "'The Repair Man Will Gyp You': Mechanics, Managers and Customers in the Automobile Repair Industry." Ph.D. dissertation, University of Missouri—Columbia.

McKay, John P. 1976. *Tramways and Trolleys: The Rise of Urban Mass Transport in Europe.* Princeton, N.J.: Princeton University Press.

McLaughlin, Charles C. 1965. "The Stanley Steamer: A Study in Unsuccessful Innovation." In *Explorations in Enterprise,* edited by Hugh Aitken, 259–72. Cambridge: Harvard University Press.

McShane, Clay. 1988. "Urban Pathways: The Street and Highway, 1900–1940." In *Technology and the Rise of the Networked City in Europe and America,* edited by Joel A. Tarr and Gabriel Dupuy, 67–87. Philadelphia: Temple University Press.

———. 1994. *Down the Asphalt Path: The Automobile and the American City.* New York: Columbia University Press.

Mason, Philip P. 1957. "The League of American Wheelmen and the Good Roads Movement, 1880–1905." Ph.D. dissertation, University of Michigan.

Maxim, Hiram Percy. 1962 [1936]. *Horseless Carriage Days.* New York: Dover.

Meyer, Perrin. 1994. Bi-logistic Growth. *Technological Forecasting and Social Change* 47: 89–102.

Miller, Spencer. 1950. "History of the Modern Highway in the United States." In *Highways in Our National Life,* edited by Jean Labatut and Wheaton J. Lane, 88–119. Princeton, N.J.: Princeton University Press.

Mokyr, Joel. 1990. *The Lever of Riches: Technological Creativity and Economic Progress.* New York: Oxford University Press.

Moline, Norman T. 1971. *Mobility and the Small Town, 1900–1930: Transportation Change in Oregon, Illinois.* Chicago: University of Chicago, Department of Geography.

Molloy, Scott. 1996. *Trolley Wars: Streetcar Workers on the Line.* Washington, D.C.. Smithsonian Institution Press.

Mom, Gilbert. 1995. "As Reliable As a Streetcar: European versus American Experiences with Early Electric Cars (1895–1914)." Paper presented at a conference of the Society for the History of Technology, Charlottesville, Va., October.

———. 1996a. "Das Holzbrettchen in der schwarzen Kiste: Die Entwicklung des Elektromobilakkumulators bei und aus der Sicht der Accumulatorenfabrik AG (AFA) von 1902–1910" [The wooden thin plate in the black box: The development of the electric accumulator at and from the viewpoint of the AFA]. *Technikgeschichte* 63, no. 2: 119–51.

———. 1996b. "Electric Vehicles: Male or Female?" Paper presented at the Intercontinental Mobile Sources Clean Air Conference, Munich, March.

———. 1997. "The Technical Dilemma: Automotive Technology between the Petrol and the Electrical Vehicle (1880–1925)." Ph.D. dissertation, Technical University of Eindhoven.

———. 1998. "Inventing the Miracle Battery: Thomas Edison and the Electric Vehicle." Unpublished paper.

Moore, Geoffrey. 1991. *Crossing the Chasm: Marketing and Selling Technology Products to Mainstream Customers.* New York: HarperBusiness.

Nader, Ralph. 1972. *Unsafe at Any Speed: The Designed-In Dangers of the American Automobile.* New York: Grossman.

Nelson, Richard, and Sidney Winter. 1982. *An Evolutionary Theory of Economic Change.* Cambridge: Belknap Press of Harvard University Press.

Nriagu, Jerome O. 1990. "The Rise and Fall of Leaded Gasoline." *Science of the Total Environment* 92 (March): 13–28.

———. 1996. "A History of Global Metal Pollution." *Science* 272 (April 12): 223–24.

Nye, David E. 1990. *Electrifying America: Social Meanings of a New Technology, 1880–1940.* Cambridge: MIT Press.

Organization for Economic Cooperation and Development. 1992. *The Urban Electric Vehicle: Policy Options, Technology Trends, and Market Prospects.* Washington, D.C.: Organization for Economic Cooperation and Development.

Owen, Wilfred. 1956. *The Metropolitan Transportation Problem.* Washington, D.C.: Brookings Institution.

Passer, Harold C. 1953. *The Electrical Manufacturers, 1875–1900.* Cambridge: Harvard University Press.

Paulsson, Martin. 1994. *The Social Anxieties of Progressive Reform: Atlantic City, 1854–1920.* New York: New York University Press.

Pfaffenberger, Bryan. 1990. "The Harsh Facts of Hydraulics: Technology and Society in Sri Lanka's Colonization Schemes." *Technology and Culture* 31, no. 3: 361–97.

———. 1992. "Technological Dramas." *Science, Technology, and Human Values* 17, no. 3: 282–312.

Pitt, Leonard, and Dale Pitt, eds. 1997. *Los Angeles A to Z: Encyclopedia of the City and County.* Los Angeles: University of California Press.

Platt, Harold L. 1991. *The Electric City: Energy and the Growth of the Chicago Area, 1880–1930.* Chicago: University of Chicago Press.

Pratt, Gil Andrews. 1992. "EVs: On the Road Again." *Technology Review* 95 (August–September): 50–59.

Preston, Howard L. 1979. *Automobile Age Atlanta: The Making of a Southern Metropolis, 1900–1935.* Athens: University of Georgia Press.

Quandt, C. O. 1995. "Manufacturing the Electric Vehicle: A Window of Technological Opportunity for Southern California." *Environment and Planning A* 27, no. 6: 835–62.

Raburn, Robert. 1988. "Motor Freight and Urban Morphogenesis with Reference to California and the West." Ph.D. dissertation, University of California, Berkeley.

Rae, John B. 1955. "The Electric Vehicle Company: A Monopoly that Missed." *Business History Review* 29 (December): 298–311.

———. 1959. *American Automobile Manufacturers: The First Forty Years.* Philadelphia: Chilton.

———. 1984. *The American Automobile Industry.* Boston: Twayne.

————. 1989. "Albert Augustus Pope." In *The Automobile Industry, 1896–1920*, edited by George S. May, 397. Volume 1 of *Encyclopedia of American Business History and Biography*. New York: Facts on File.

Rajashekara, Kaushik. 1994. "History of Electric Vehicles in General Motors." *IEEE Transactions on Industry Applications* 30, no. 4: 897–904.

Reese, Neal J., and James M. Olson, eds. 1995. *What Might Have Been: The Social Psychology of Counterfactual Thinking*. Mahwah, N.J.: Erlbaum.

Richards, Deanna J., Braden R. Allenby, and Robert A. Frosch. 1994. "The Greening of Industrial Ecosystems: Overview and Perspective." In *The Greening of Industrial Ecosystems*, edited by Braden R. Allenby and Deanna J. Richards, 1–22. Washington, D.C.: National Academy Press.

Romney, George. 1950. "The Motor Vehicle and the Highway: Some Historical Implications." In *Highways in Our National Life*, edited by Jean Labatut and Wheaton J. Lane. Princeton, N.J.: Princeton University Press.

Rose, Mark H. 1990 [1979]. *Interstate: Express Highway Politics, 1939–1990*. Knoxville: University of Tennessee Press.

Rosenberg, Nathan. 1976. *Perspectives on Technology*. New York: Cambridge University Press.

————. 1982. *Inside the Black Box: Technology and Economics*. New York: Cambridge University Press.

Roth, Matt. 1997. "Mulholland Highway and the Engineering Culture of Los Angeles in the 1920s." Paper presented at the annual meeting of the Society for the History of Technology, Pasadena, Calif., October.

St. Clair, David J. 1986. *The Motorization of American Cities*. New York: Praeger.

Schallenberg, Richard. 1982. *Bottled Energy: Electrical Engineering and the Evolution of Chemical Energy Storage*. Philadelphia: American Philosophical Society.

Scharff, Virginia. 1991. *Taking the Wheel: Women and the Coming of the Motor Age*. New York: Free Press.

————. 1992. Gender, Electricity, and Automobility. In *The Car and The City: The Automobile, the Built Environment, and Daily Urban Life*, edited by Martin Wachs and Margaret Crawford, 75–85. Los Angeles: University of California Press.

Schatzberg, Eric. 1993. "The Mechanization of Urban Transit in the United States." In *Technological Competitiveness: Contemporary and Historical Perspectives on the Electrical, Electronics, and Computer Industries*, ed. William Aspray, 225–42. New York: IEEE Press.

Schiffer, Michael B. 1994. *Taking Charge: The Electric Automobile in America*. Washington, D.C.: Smithsonian Institution Press.

————. 1995. "Social Theory and History in Behavioral Archaeology." In *Expanding Archaeology*, edited by James M. Skibo, William H. Walker, and Axel E. Nielsen, 22–35. Salt Lake City: University of Utah Press.

Schipper, Lee. 1992. "Electric Vehicles in a Broader Context: Too Early or Too Late?" In *The Urban Electric Vehicle: Policy Options, Technology Trends and Market Prospects*, 381–85. Stockholm: Organization for Economic Cooperation and Development.

Schivelbusch, Wolfgang. 1986 [1977]. *The Railway Journey: The Industrialization of Time and Space in the 19th Century*. Berkeley: University of California Press.

————. 1988 [1983]. *Disenchanted Night: The Industrialization of Light in the Nineteenth Century*, translated by Angela Davies. Berkeley: University of California Press.

Schot, Johan W. 1992. "Constructive Technology Assessment and Technology Dynamics: The Case of Clean Technologies." *Science, Technology, and Human Values* 17, no. 1: 36–56.

Schroeder, Fred E. H. 1986. "More 'Small Things Forgotten': Domestic Electrical Plugs and Receptacles, 1881–1931." *Technology and Culture* 27 (July): 525–43.

Scott, Allen J. 1993. *Electric Vehicle Manufacturing in Southern California: Current Developments, Future Prospects*. Los Angeles: University of California, Los Angeles; Lewis Center for Regional Policy Studies.

Seely, Bruce E. 1987. *Building the American Highway System: Engineers As Policy Makers*. Philadelphia: Temple University Press.

————. 1988. "Diffusion of Science into Engineering: Highway Research at the Bureau of Public Roads, 1900–1940." In *Transfer and Transformation of Ideas and Material Culture*, edited by Peter J. Hugill and D. Bruce Dickson, 143–62. College Station: Texas A&M University Press.

Shnayerson, Michael. 1996. *The Car That Could: The Inside Story of GM's Revolutionary Electric Vehicle*. New York: Random House.

Sicilia, David B. 1990. "Selling Power: Marketing and Monopoly at Boston Edison, 1886–1929." Ph.D. dissertation, Brandeis University.

———. 1991. "Selling Power: Marketing and Monopoly at Boston Edison, 1886–1926." *Business and Economic History* 20, 2d series: 27–31.

Smith, Merrit Roe, and Leo Marx, eds. 1994. *Does Technology Drive History: The Dilemma of Technological Determinism*. Cambridge: MIT Press.

Snell, Bradford C. 1974. *American Ground Transport*. Washington, D.C.: U.S. Senate, Subcommittee on Antitrust and Monopoly of the Committee on Judiciary.

Staudenmaier, John M. 1985. *Technology's Storytellers: Reweaving the Human Fabric*. Cambridge: MIT Press/Society for the History of Technology.

Sullivan, Joseph P. 1995. "From Municipal Ownership to Regulation: Municipal Utility Reformers in New York City, 1880–1907." Ph.D. dissertation, Rutgers University.

Taylor, Barbara E. 1995. *The Lost Cord: A Storyteller's History of the Electric Car*. Columbus, Ohio: Greydon.

Tenner, Edward. 1996. *Why Things Bite Back: Technology and the Revenge of Unintended Consequences*. New York: Knopf.

Tetlock, Philip E., and Aaron Belkin, eds. 1996. *Counterfactual Thought Experiments in World Politics*. Princeton, N.J.: Princeton University Press.

Thompson, F. M. L. 1983. *Horses in European Economic History: A Preliminary Canter*. London: British Agricultural Society.

Thompson, George V. 1954. "Intercompany Technical Standardization in the Early American Automobile Industry." *Journal of Economic History* 14, no. 1: 1–20.

U.S. Bureau of the Census. 1917. *Abstract of the Census of Manufactures, 1914*. Washington, D.C.: Government Printing Office.

———. *Fourteenth Census, 1920*. Volume 10: *Manufactures*. Washington, D.C.: Government Printing Office.

U.S. Department of Commerce, Panel on Electrically Powered Vehicles. 1967. *The Automobile and Air Pollution*. Washington, D.C.: Government Printing Office.

U.S. Department of Energy, Energy Research and Development Administration. 1977. *Electric Vehicle Systems, FY 1978*. EDP/C-01. Washington, D.C.: Government Printing Office, August.

U.S. Department of Energy. 1980. *Near Term Electric Test Vehicle: Phase II—Final Report*. DOE/CS/512–94–01. Washington, D.C.: Government Printing Office, October.

U.S. Department of Transportation, Bureau of Transportation Statistics. 1995. *National Transportation Statistics, 1996*. Washington, D.C.: Government Printing Office.

Volti, Rudi. 1990. "Why Internal Combustion?" *Invention and Technology* 6: 42–47.

von Hippel, Eric. 1986. "Lead Users: A Source of Novel Product Concepts." *Management Science* 32, no. 7 (July): 791–805.

Wachs, Martin, and Margaret Crawford, eds. 1992. *The Car and the City*. Los Angeles: University of California Press.

Wager, Richard. 1986. *Golden Wheels: The Story of the Automobiles Made in Cleveland and Northeastern Ohio, 1892–1932*. Cleveland: Western Reserve Historical Society.

Wakefield, Ernest H. 1994. *History of the Electric Automobile: Battery-Only Powered Cars*. Warrendale, Pa.: Society of Automotive Engineers.

Waldrop, M. Mitchell. 1992. *Complexity: The Science at the Edge of Order and Chaos*. New York: Simon and Schuster.

Watson, Ann Y., Richard R. Bates, and Donald Kennedy, eds. 1988. *Air Pollution, the Automobile, and Public Health*. Washington, D.C.: National Academy Press.

Woodbury, George. 1950. *The Story of a Stanley Steamer*. New York: Norton.

Wouk, Victor. 1995. "Hybrids: Then and Now." *IEEE Spectrum* 32 (July): 16–21.

Wren, Genevieve. 1989. "Pedro G. Salom." In *The Automobile Industry, 1896–1920*, edited by George S. May, 408–10. Volume 1 of *Encyclopedia of American Business History and Biography*. New York: Facts on File.

Yergin, Daniel. 1991. *The Prize: The Epic Quest for Oil, Money and Power*. New York: Simon and Schuster.

Young, Charles Frederic T. 1860. *The Economy of Steam Power on Common Roads*. London: Atchley.

Zorpette, Glenn. 1997. "Alan Cocconi: Electric Cars and Pterosaurs Are My Business." *Scientific American* 276 (May): 32–33.

Index

advertising campaign, 101
alkaline battery, 127, 181, 196, 197, 200–201
Allenby, Braden, 233–234
alternative transport technology: assessment of, 79–84, 243n11; government support for, 234–236; internal combustion and, 13, 19–21, 236–237
ambulances, 217–218
America Adopts the Automobile, 1895–1910 (Flink), 18
American Railway Express Company, 150
American Street Railway Association, 126
Anderson, William C., 98, 121, 258n98
Armstrong, Frank C., 46, 67, 68, 71, 248n42
Atkins, W. H., 93
Atlantic City (N.J.), electric cabs in, 67–71
Atwood, William Hooker, 41, 57
Automobile Club of America, 184, 215, 216
Automobile Club of France, 59
automobile industry: development of, 12–13; historiography of, 13, 15, 20–21; legislation regarding, 17; transportation system and, 4–5
automobile manufacture. *See* manufacturers

"Automobiles versus Trolley Cars," 80–81
automobile system, 12–25, 217–223, 226; competition in, 22–24; dynamics of, 187–188; electric alternatives for, 13, 19–21; electrification of, 221–223; environmental costs of, 23–24, 232–233; social context of, 14–15, 25; suburbanization and, 18, 153; technology of, 14, 232–233, 234; triumph of internal combustion in, 15, 16–17; usage patterns of, 217–218

Babcock, F. A., 184, 188
Bailey, E.W.M., 167–168, 169, 180, 187, 188; and Bailey roadster (electric vehicle), 167, 202
Baker, Day, 105, 143, 238, 239, 245n9
Baker, Walter, and Baker Motor Vehicle Company, 125, 141, 172, 184
Bartlett, E. J., 139, 180
Bartlett, J. C., 112
batteries, 24–25, 99, 189; design of, 92, 106; Edison alkaline, 127, 181, 196, 197, 200–201; lead-acid, 197, 198, 200, 206, 207; maintenance of, 69–70, 247n41, 258n90; owner knowledge of, 176–177; problems with, 45–46, 59–60, 196; standardization of, 104–105; systems for, 33–34, 66, 196, 197–198;

About the Author

David A. Kirsch is a visiting assistant professor and AT&T Faculty Fellow in Industrial Ecology at the Anderson Graduate School of Management at the University of California, Los Angeles, where he teaches business history, business strategy, and environmental management. His research examines the long-run dynamics of technological change and industrial evolution with specific emphasis on the problems of competing technologies, institutions, and standards in shaping the outcomes of technological battles and the implications of these battles for social and environmental change. He received his Ph.D. in the history of technology from Stanford University.